Automobile Automation

Automobile Automation
Distributed Cognition on the Road

Victoria A. Banks and Neville A. Stanton

CRC Press
Taylor & Francis Group
Boca Raton London New York

CRC Press is an imprint of the
Taylor & Francis Group, an **informa** business

CRC Press
Taylor & Francis Group
6000 Broken Sound Parkway NW, Suite 300
Boca Raton, FL 33487-2742

International Standard Book Number-13: 978-1-138-19683-4 (Paperback)
978-1-138-06793-6 (Hardback)

Library of Congress Cataloging-in-Publication Data

Names: Banks, Victoria A., author. | Stanton, Neville A. (Neville Anthony),
1960- author.
Title: Automobile automation : distributed cognition on the road / Victoria
A. Banks, Neville A. Stanton.
Description: Boca Raton : CRC Press, 2017. | Includes bibliographical
references and index.
Identifiers: LCCN 2017005006| ISBN 9781138196834 (pbk. : acid-free paper) |
ISBN 9781315295657 (ebook)
Subjects: LCSH: Automobiles--Automatic control. |
Automobile driving--Human factors.
Classification: LCC TL152.8 .B36 2017 | DDC 629.2/72--dc23
LC record available at https://lccn.loc.gov/2017005006

Visit the Taylor & Francis Web site at
http://www.taylorandfrancis.com

and the CRC Press Web site at
http://www.crcpress.com

Transportation Human Factors:
Aerospace, Aviation, Maritime, Rail, and Road Series

Series Editor
Professor Neville A Stanton
University of Southampton, UK

Automobile Automation: Distributed Cognition on the Road
Victoria A. Banks, Neville A. Stanton

Eco-Driving: From Strategies to Interfaces
Rich C. McIlroy, Neville A. Stanton

For Mark and Caitlin

Vicky

For Maggie, Josh and Jem

Neville

Contents

Preface

This book came about through our work on applying the ideas of Distributed Cognition to automobile automation. We have shown that some of the cognitive functions traditionally performed by the human driver of manually controlled vehicles are going to be performed by automation. The dynamic nature of driving means that cognitive functions (such as Monitor, Anticipate, Detect, Recognise, Decide, Select and Respond) change momentarily, in light of changes in the task, environment and interactions with other road users. In our research, we have shown how these cognitive functions may be allocated, dynamically, to different agents in the vehicle (both human and technological). We have undertaken both modelling and empirical work in a cycle of model-test-model in order to predict the performance of automated systems and validate the modelling work. To this end, we have used a variety of Human Factors methods to show how the driver may be incorporated into engineering analysis of future technologies. We are extremely grateful to our colleagues at Jaguar Land Rover, who have presented us with the design challenges and facilitated the simulator, test-track and on-road studies. The insights they have provided us with of future automotive systems have been invaluable.

This book may be used in several ways. As a primer for the Human Factors issues in automobile automation, it can bring the reader up-to-speed on the issues and approaches, as well as providing empirical evidence on the range of behaviours in automated vehicles. The book also presents methods that can be used in the different stages of design, from formative approaches for modelling initial concepts to summative approaches for evaluation of technologies in simulators, on test-tracks and on the road. For the researcher we offer studies, concepts and ideas to stimulate further work. For the practitioner, we offer a review of the field, data from studies and indications of where the future lies. We have performed one of the first ever studies of vehicle automation on the road, looking at driver behaviour with a lane change system. Our pedigree of conducting research into vehicle automation goes back to the early 1990s, when Neville was one of the very few researchers undertaking studies into vehicle automation.

There can be no doubt that road vehicle automation will be with a common feature very soon. Tesla's autopilot system offers early insight into the ways in which it can be deployed. Certainly, the Tesla system has stimulated technological progress. BMW, Jaguar, Mercedes, Volvo and other vehicle manufacturers are not far behind. Predictions about other companies product launches are being made for around 2020. We hope our book helps designers and engineers consider the role of the driver in the future of vehicle automation. We have expressed concerns about bringing the driver back into the vehicle control loop in a controlled manner, by ensuring the driver has timely and salient information when in supervisory mode. We have also extended our work to consider the macro-level transport system concerns of mixed levels of automation on the road. All of these issues need to be resolved if vehicle automation is to contribute to road safety in the future.

Acknowledgements

We would like to thank the Engineering and Physical Sciences Research Council (EPSRC) and Jaguar Land Rover for funding this work.

Special thanks goes to Jim O'Donoghue for all of the support you have provided in making much of this research possible.

Authors

Dr. Victoria A. Banks is a postgraduate research fellow in the Human Factors Engineering Team, which is part of the Transportation Research Group at the University of Southampton. She was recently awarded her Engineering Doctorate by the University of Southampton in 2016 and also holds a BSc Psychology (Hons) award. Her research interests include modelling, analysing and evaluating the Human Factors implications of increasing levels of autonomy on driver behaviour. She has published over 10 articles related to the field of automobile automation and, in 2014, Dr. Banks was invited to speak at the Transport Select Committee Event at Southampton University, from which she was invited to write an article for ITS International. Dr. Banks has previously worked with one of the largest vehicle manufacturers in the United Kingdom and was involved with the design and development of future automated technologies.

Professor Neville A. Stanton, PhD, DSc, is a chartered psychologist, chartered ergonomist and chartered engineer. He holds the Chair in Human Factors Engineering in the Faculty of Engineering and the Environment at the University of Southampton in the United Kingdom. He has degrees in Occupational Psychology, Applied Psychology and Human Factors Engineering and has worked at the Universities of Aston, Brunel, Cornell and MIT. His research interests include modelling, predicting, analysing and evaluating human performance in systems as well as designing the interfaces and interaction between humans and technology. Professor Stanton has worked on design of automobiles, aircraft, ships and control rooms over the past 30 years, on a variety of automation projects. He has published 35 books and over 270 journal papers on Ergonomics and Human Factors. In 1998, he was presented with the Institution of Electrical Engineers Divisional Premium Award for research into System Safety. The Institute of Ergonomics and Human Factors in the United Kingdom awarded him The Otto Edholm Medal in 2001, The President's Medal in 2008 and The Sir Frederic Bartlett Medal in 2012 were awarded for his contributions to basic and applied ergonomics research. The Royal Aeronautical Society awarded him and his colleagues the Hodgson Prize in 2006 for research on design-induced, flight-deck, error published in *The Aeronautical Journal*.

The University of Southampton awarded him a Doctor of Science in 2014 for his sustained contribution to the development and validation of Human Factors methods.

List of Abbreviations

ACC	Adaptive Cruise Control
ADAS	Advanced Driver Assistance System
AEB	Autonomous Emergency Brake
AGNA	Applied Graphic and Network Analyses
BASt	Bundesanstalt für Straßenwesen
CAV	Connected and Autonomous Vehicle
CDM	Critical Decision Method
DARPA	Defence Advanced Research Project Agency
DD	Driver Driving
DM	Driver Monitor
DND	Driver Not Driving
DSA	Distributed Situation Awareness
DSRC	Dedicated Short-Range Communication
DSSQ	Dundee Stress State Questionnaire
EA	External Agency
EAST	Event Analysis of Systemic Teamwork
ERTRAC	European Road Transport Research Advisory Council
Euro NCAP	European New Car Assessment Programme
FAA	Federal Aviation Administration
GPS	Global Satellite Positioning
HMI	Human–Machine Interface
IPSGA	Information, Position, Speed, Gear and Acceleration
LDWS	Lane Departure Warning System
LIDAR	Light Detection and Ranging
MART	Malleable Attentional Resource Theory
NASA-TLX	National Aeronautics and Space Administration – Task Load Index
NHTSA	National Highway Traffic Safety Adminstration
OECD	Organisation for Economic Cooperation and Development
OSD	Operator Sequence Diagram
PF	Pilot Flying
PM	Pilot Monitor
PNF	Pilot Not Flying
RE	Road Environment
SA	Situation Awareness
SAE	Society of Automotive Engineers
SAGAT	Situation Awareness Global Assessment Technique
TA	Tense Arousal
THW	Time Headway
TMC	Traffic Management Centre
UTMC	Urban Traffic Management Control

V2I	Vehicle-to-Infrastructure
V2V	Vehicle-to-Vehicle
VMS	Variable Message Sign
VPA	Verbal Protocol Analysis
WHO	World Health Organisation

List of Figures

List of Tables

1 Introduction to Automobile Automation

INTRODUCTION

The Defence Advanced Research Project Agency (DARPA) has fuelled interest into the field of 'vehicle automation' since the 1950s as it was recognised that automated vehicles could be used to gather intelligence, be used in surveillance operations and for target acquisition and reconnaissance (Rouff and Hinchey, 2012). The agency placed emphasis on maintaining technological superiority and security as well as reducing the number of personnel required on the ground.

DARPA is most famously recognised for its Grand Challenges (2004, 2005) and Urban Challenge (2007) that invited teams to build and design fully autonomous vehicles. The first Grand Challenge in 2004 aimed to show that autonomous vehicles could undertake resupply missions in unfamiliar desert terrain. Although no vehicles completed the course, success was finally achieved in 2005, demonstrating that unmanned vehicles could navigate across remote environments, on a variety of road surfaces with different obstacles and with limited or no global positioning satellites (Rouff and Hinchey, 2012). The 2007 Urban Challenge was designed to test the ability of autonomous vehicles to navigate safely and effectively through populated areas to simulate supply missions while adhering to normal driving laws. At this point, safety was of utmost importance and all vehicles had to be equipped with a form of 'E-Stop' – autonomous emergency braking – to maintain the safety of DARPA employees and spectators. It is these advancements that fuelled research and innovation within the automotive industry as the capabilities of automated vehicles to improve the safety of the roads and its occupants had been recognised.

In line with the advancements facilitated by DARPA, the introduction of automated driving features into 'civilian' life has gradually risen since 2000. The main purpose of automated driving features from a marketing point of view is to continue the trend of safe, comfortable, efficient and enjoyable personal travel as well as bring about improvements to traffic efficiency and fuel consumption (e.g. Khan et al., 2012; Ward, 2000). In the 1990s, cruise assist technologies increased in popularity, and as autonomy appeared to improve the driving experience, more advanced features were developed.

While fully autonomous vehicles (i.e. vehicles requiring no human operator) were developed for the DARPA Challenges and more recently by Google in 2014, automation within the automotive industry requires an acknowledgement of Human Factors in the design of automated driving features because the driver remains an active participant within the driving task to some extent. Although over recent years, technological advancements have meant that vehicles have become increasingly capable of performing the same functions as the driver to a much greater degree, there continues to be a stipulation within the law that drivers must remain in overall control of their vehicle (e.g. Article 8 of the Vienna Convention, 1968). A recent amendment to the convention in 2014 (introduced in 2016) states that driverless cars are allowed on the road as long as they can be overridden by a human driver. This means now more than ever, the driver needs to remain capable of regaining control of an automated vehicle and be supported to do so following prolonged exposure to periods of highly automated driving where boredom and fatigue may become increasingly problematic.

The research presented in this book offers one of the first acknowledgements of how the introduction of automation into the driving task fundamentally changes the role of the driver within it using task analysis modelling techniques. The aim therefore is not to deliver specific data or guidelines about 'how' to manage a transfer of control in autonomy but instead identify and increase our understanding of the *changing* role of the driver within the totality of the driving system.

Safety research suggests that driver inattentiveness and lack of timely response to unpredictable or incomplete information are the most common driver errors that result in vehicular accidents (Amditis et al., 2010; Cantin et al., 2009; Donmez et al., 2007; Khan et al., 2012; Stanton and Salmon, 2009). These external factors are typically random events that evolve to form complex interactions between the driver and the vehicle (Khan et al., 2012). Without automated assistance, the driver may be underprepared or lack the skill required to respond to the situation accordingly. For this reason, automated vehicles have great potential to improve the safety of our roads and in turn reduce the economic burden of any cumulative effect as a result of an accident such as sick pay through injury and impact to businesses if roads are closed. To put this into focus, the World Health Organisation (WHO) has stated that if current road traffic accident trends continue, the annual fatalities as a consequence of such accidents will increase to 2.34 million by 2020 (Khan et al., 2012). In 2012, the WHO declared that approximately 1.3 million people per annum die as a result of road traffic accidents. Nearly half of these (46%) are considered to be 'vulnerable road users'. Deaths resulting from road traffic accidents are the leading cause of injury mortality, offering a clear justification for investing time into the field of vehicle automation. If the benefits of automation outweigh potential costs, then automation may prove to be beneficial in economic, societal and environmental terms (Khan et al., 2012; Stanton and Marsden, 1996; Young et al., 2011). However, despite the expectation that automation will bring about enhancement of road safety, such hypotheses require further validation (Stanton and Marsden, 1996). Further research is needed to assess the degree to which automation can reduce the overall number of driver errors that are often implicated as the cause of many vehicular accidents.

Since 1997, the European New Car Assessment Programme (Euro NCAP) has continued to encourage vehicle manufacturers to exceed the minimal safety requirements that are required by law. It also aims to ensure that stringent guidelines and testing protocols are rigorously enforced to ensure that potential new customers are given transparent safety information through use of its internationally recognised Five-Star Rating Scheme. By rewarding technologies, Euro NCAP pushes vehicle manufactures to accelerate their standard fitment of key automated safety technologies such as Blind Spot Monitoring, Lane Support Systems, Speed Alert Systems,

Autonomous Emergency Braking, Automatic Emergency Call and Pre-Crash Systems.

There are of course other reasons why automation may be beneficial. For example, automated driving may not only improve road safety, but also reduce traffic congestion, exhaust gas emissions and fuel consumption according to the European Commission (2011). Interest in automated driving as a form of 'Traffic Management System' continues to grow as demonstrated through the 9th Intelligent Transport Systems European Congress (2013), which included a special interest session that looked specifically at the future of highly automated vehicles (including highway trucks and vehicle platooning) as well as automated urban transportation. Although air quality has been an environmental concern for some time, transport is currently a major source of air pollution within the United Kingdom, and with car use set to increase further, more needs to be done to tackle the problem of congestion and its associated impacts both economically and environmentally (Fagnant and Kockelman, 2015). There are a number of approaches that can be used to improve air quality, and new vehicle technologies can play an important role in addressing these environmental issues further.

With systems design plagued by criticism for failing to adequately define the role of the human operator within a system, there is concern among the Ergonomics and Human Factors community that automated subsystems in driving may create more problems than they solve. Failing to acknowledge the role of the driver in an automated driving system therefore may lead to undesirable behavioural adaptation as a result of inadequately anticipating the changing role of the driver within the system. This is likely to become even more problematic as multiple vehicle subsystems, operating at different levels of automation, are interacting. It is also a very important area of study given recent legislation that requires the driver to be capable of regaining control of an automated vehicle.

This research attempts to address concerns surrounding driver behavioural adaptation in three main ways:

1. Increase the awareness of Human Factors in the design of automated aids by focussing on the interaction that occurs between the driver and other system agents

    With growing concern that the role of the driver is not being fully recognised in the design of automated driving systems, it is important to focus upon the interaction that occurs between the driver and system agents at differing levels of autonomy. This allows for exploration of the diminishing role of the driver with regard to direct vehicle control and what this may mean to the overall functioning of this sociotechnical system (Walker et al., 2010). Importantly, this book is primarily concerned with understanding the role of the driver at intermediate levels of automation. Thus, it is not a book about 'driverless' vehicles whereby human operators are free to do whatever they want. The authors argue that no matter how small their role, the role of the driver remains an important design consideration.

2. Assess the appropriateness of automation deployment and context of use

    Human Factors would argue that even though it may be possible to fully automate a vehicle, it may not always be appropriate to do so given the limits of human attention needed to execute a required response. An automatic braking system, for example, could relinquish driver control over a critical safety function. This may be appropriate to do so in scenarios whereby the driver has not got the capacity to respond, such as 500 ms prior to a collision. Such an autonomous feature, however, may cause drivers to become more reliant on its presence. This may result in increased reaction times and stopping distances as drivers 'wait' for the system to engage.

3. Provide design guidance on automated features based upon experimental evidence

Being able to provide systems design guidance to vehicle manufacturers is extremely important to ensure that the functionality of driver–vehicle interaction is optimised as far as is reasonably practicable.

OUTLINE OF BOOK

CHAPTER 1: INTRODUCTION TO THIS BOOK

This initial chapter introduces the area of driving automation and outlines the aims and objectives of the research. It also includes a summary of each chapter and indicates the contribution to knowledge.

CHAPTER 2: ON THE ROAD TO FULL VEHICLE AUTOMATION

This chapter introduces the concept of automation and the different levels at which it can be introduced into a system, thus altering the role of the human operator within it. Multiple automation taxonomies are discussed that have sought to better define 'who' is doing 'what' at varying levels of automation. What all automation taxonomies have in common is that at higher levels, the level of control that the human operator has over a system is reduced. However, this does not mean that they become completely removed from the system altogether. Instead, they remain to some extent within the control-feedback loop. This is because they continue to receive feedback from the automated system and their wider environment. In terms of driving, the driver will continue to receive feedback from the automated system via the Human–Machine Interface (HMI) within the vehicle in addition to feedback from the wider road environment even when the vehicle is capable of performing much of the driving task autonomously. This means that driver responsibilities continue to change as the level of automation increases. Assessing whether drivers are able to adhere to these changing responsibilities requires an acknowledgement of key Human Factors considerations. Chapter 2 reviews the literature relating to four key Human Factors concepts: situation awareness, driver workload, trust and skill and concludes that automation can have both positive and negative effects on each of these dimensions.

CHAPTER 3: ADOPTING A SYSTEMS ENGINEERING VIEW

Previous research into automation has traditionally been either Technology-Focussed or Human-Centred. However, this chapter adopts an increasingly popular sociotechnical view that takes into consideration both the strengths and the weaknesses of all system agents (both human and non-human). With the allocation of system function being key to understanding how automation may affect the role of the driver within the system network, Chapter 3 introduces the concept of Distributed Cognition (Hutchins, 1995a). This paradigm aims to better define how tasks can be partitioned between system agents. The application of Distributed Cognition to driving is a new and unexplored medium, yet there appears to be great benefit in doing so. This is because it enables system engineers and designers to acknowledge the new role of the driver in an automated driving system. Chapter 3 introduces a two-phase Systems Design Framework that applies Distributed Cognition to driving. The first 'modelling' phase uses a well-established and popular Ergonomics technique (Operator Sequence Diagrams [OSDs]; Brooks, 1960; Kurke, 1961) to represent the interactions that take place between system agents. The second 'experimental' phase aims to validate these system models through the collection and analysis of empirical data. Chapter 3 demonstrates the first phase of this framework using an example of Pedestrian Autonomous Emergency Brake (AEB).

CHAPTER 4: EXPLORING THE USE OF VERBAL PROTOCOL ANALYSIS AS A TOOL TO ANALYSE DRIVER BEHAVIOUR

Although the representations that are afforded by system modelling provide an insight into the behaviour and interaction that occur between multiple system agents, they are unable to represent the underlying cognitive behaviour of the driver. Chapter 4 explores the use of Verbal Protocol Analysis (VPA; Ericsson and Simon, 1993) as a tool to both validate and extend the visual representations of automated driving systems. VPA is a direct observation method that can capture the underlying processes that mediate behavioural outcomes that are often represented by hard data alone. In this case, hard data were supplemented by the analysis of driver verbalisations relating to driving emergencies experienced using the Southampton University Driving Simulator. The study in Chapter 4 was a pilot study that resulted in a number of practical recommendations for future research being put forward. These recommendations contribute to a methodological advance in using retrospective verbal protocols and contributed to the design of the investigation discussed in Chapter 5.

CHAPTER 5: USING RETROSPECTIVE VERBAL PROTOCOLS TO EXPLORE DRIVER BEHAVIOUR IN EMERGENCIES

This chapter introduces models of Driver Decision-Making in Emergencies (DDMiE) to investigate how the level and type of automation (specifically in relation to systems design) may affect driver decision-making and subsequent responses to critical braking events. Network analysis was used to interrogate retrospective verbalisations, making it possible to quantitatively analyse driver decision-making processing. Findings suggest that while automation does not alter the decision-making pathway (e.g. the processes between hazard detection and response remain similar), it does appear to significantly weaken the links between information-processing nodes. This reflects an unintended yet emergent property within the task network that could mean that we may not be improving safety in the way we expect.

Chapter 6: The Effect of Systems Design on Driver Behaviour

Chapter 6 builds upon the work presented in Chapter 5 by analysing the performance data generated by the Southampton University Driving Simulator during the same study in light of evidence within driver verbalisations. Data were analysed with a view to assess the appropriateness of systems design and context of use at varying levels of automation within driving emergencies. This was based upon the suggestion that a 'silent' and 'invisible' AEB system would be less likely to lead to negative behavioural adaptation on behalf of the driver. Chapter 6 explores whether this was actually the case. Despite significant improvements in road safety, the data suggested that systems design had a direct effect on driver–vehicle interaction patterns with drivers being more likely to relinquish control of the braking effort to a warning-based system of AEB than a non-warning-based system of AEB. This means that while we may be improving the safety of other road users, we may not always improve the safety of our drivers.

Chapter 7: What Is Next for Vehicle Automation? From Design Concept through to Prototype

To the average driver, the concept of automation in driving infers that they can become completely 'hands and feet free'. This is a common misconception however, one that has been shown through the application of network analysis to new Cruise Assist technologies that began to enter the marketplace in 2016. This chapter introduces the concept of Driver-Initiated Automation, an approach that appears to be the implementation route of choice for next-generation automated highway features (e.g. Tesla's Autopilot feature). Chapter 7 uses Phase 1 of the Systems Design Framework introduced in Chapter 3 to show how the role of the driver remains an integral part of the driving system using Distributed Cognition. This implicates the need for designers to ensure that drivers are provided with the tools necessary to remain actively in-the-loop despite being given increasing opportunities to delegate their control to the automated subsystems.

Chapter 8: Discovering Driver–Vehicle Coordination Problems in Early-Stage System Development

Chapter 8 discusses a case study that was designed to investigate possible functionality issues of a Driver-Initiated Command and Control System of Automation. Verbalisations and subjective reports of mental workload and stress revealed evidence of different driver–vehicle coordination problems (i.e. mode confusion and automation surprise) depending upon the level of driver familiarity with the system.

CHAPTER 9: DRIVER-INITIATED DESIGN: AN APPROACH TO KEEPING THE DRIVER IN CONTROL?

Automated automobiles will be on our roads within the next decade, but the role of the driver has not yet been formerly recognised or designed. Rather, drivers are often left in a passive monitoring role until they are required to reclaim control from the vehicle. Chapter 9 discusses a study that tested the idea of Driver-Initiated Automation, in which the automation offered decision support related to an automatic overtake that could be either accepted or ignored. Despite putting drivers in control of the automated system by enabling them to accept or ignore behavioural suggestions (e.g. overtake), there were still issues associated with increased workload and decreased trust. These issues are likely to have arisen due to the way in which the automated system was designed. Recommendations for improvements to systems design were made that sought to improve ratings of trust and make the role of the driver, with regards to their authority over the automated system, more transparent.

CHAPTER 10: DISTRIBUTED COGNITION IN THE ROAD TRANSPORTATION NETWORK: A COMPARISON OF 'CURRENT' AND 'FUTURE' NETWORKS

Unlike previous chapters, Chapter 10 recognises that the functioning of the overall road transportation network is based upon an infinite number of complex interactions and interdependencies between multiple system agents at a number of levels. Chapter 10 demonstrates how aspects of the Systems Design Framework can be used to explore Distributed Cognition at a macro-level. Comparisons are made between the conventional transportation network and one of the future that sees increasing levels of connected and autonomous vehicles (CAVs) on the road. Risks associated with network resilience are explored.

CHAPTER 11: SUMMARY OF FINDINGS AND RESEARCH APPROACH

The final chapter summarises the research objectives in light of the findings presented in this book and considers the contributions made to knowledge. An evaluation of the research approach highlights that while qualitative research methodologies are often criticised for their lack of objectivity, they provide researchers with an insight into 'how' and 'why' drivers use automation in the way that they do. Relying upon quantitative data alone would have resulted in an incomplete representation of the issues relevant to driving automation. Consideration is also given to the theoretical, methodological and practical implications of the research based upon the development of the Systems Design Framework and the tools it uses in extending our understanding of the role of the driver within an automated driving system.

To sum, the work presented in this book contributes to our understanding of Distributed Cognition (Hutchins, 1995a) in driving. While traditionally, descriptions of Distributed Cognition have been solely narrative, this book has further developed our understanding of the allocation of system function by modelling Distributed Cognition using well-established Ergonomics techniques. This book proposes that a comprehensive understanding of Distributed Cognition in driving can be achieved by following a six-step Systems Design Framework that gives rise to the opportunity to highlight design weaknesses or areas for consideration and provide design solutions based upon experimental data through which model validation can be achieved. The findings of this research project have not only contributed to our understanding of Distributed Cognition in driving but also provided a platform to explore and identify system design weaknesses that may have undesirable consequences on driver behaviour.

2 On the Road to Full Vehicle Automation

INTRODUCTION

The Human Factors issues pertaining to vehicle automation have been speculated about since the 1970s (Sheridan, 1970). Simply removing the driver from the control-feedback loop and eliminating their responsibility over safe vehicle operation do not warrant Human Factors completely redundant. Instead, vehicles operating at increased levels of autonomy with 'self-driving' capabilities open up new avenues of investigation. Fully autonomous cars promise to deliver abundant socioeconomic advantages (Casner et al., 2016) including improvements to traffic flow, mobility and significantly improved road safety. With 90% of accidents being attributable to driver error (Smiley and Brookhuis, 1987; Stanton and Salmon, 2009), vehicle manufacturers may have good reason to remove drivers from active control. There are of course other benefits of 'driverless' vehicles, and this is to give vehicle occupants time to engage in other tasks not related to driving (Fagnant and Kockelman, 2015). This is seen as one of the main drivers for market implementation to improve comfort and convenience. According to the Department for Transport (2015), U.K. drivers spend, on average, 235 h a year behind the wheel. This equates to approximately six working weeks whereby the driver has no spare capacity to engage in other tasks. The advent of fully automated vehicles could therefore completely transform our experience of driving and provide the driver with additional productive time (making the journey similar to taking public transport – without the drawbacks).

However, we cannot overlook the fact that individuals may want to resume control from the vehicle at some point. This means that systems designers need to ensure that the drivers are supported. Thus, increasing levels of autonomy in driving do not eliminate all of the Human Factors issues that are typically associated with lower levels. We already know from the literature that automation within the driving task can lead to decreased situation awareness (SA; Stanton and Young, 2005; Stanton et al., 2011), erratic changes to driver workload (de Winter et al., 2014, 2016; Stanton et al., 1997; Young and Stanton, 2002, 2004), skill degradation (Stanton and Marsden, 1996) and issues relating to trust (Walker et al., 2016), over-reliance and complacency (Stanton, 2015). It seems likely that some, if not all of these, will remain enduring challenges for systems designers as long as the driver remains within the control-feedback loop to some extent.

LEVELS OF AUTOMATION

A recent review of the literature by Vagia et al. (2016) states that since the 1950s, a total of 12 automation taxonomies have been developed. One of the oldest and most widely cited taxonomies was developed by Sheridan and Verplanck (1978). This comprehensive 10-level automation taxonomy specified which system functions were the responsibility of the human operator and which were the responsibility of the computer system. It ranged from fully manual control (Level 1) to fully automated control (Level 10) with intermediate levels combining differing levels of human and computer control. A later automation model by Endsley and Kaber (1999) sought to better define these intermediate levels by identifying 'who' was doing 'what' at each level of automation. The advantage of Endsley and Kaber's 10-level taxonomy is the explicit nature in which system monitoring, strategy generation, decision-making and response execution have been assigned to both human and computer or as single entities. It meant that it had wider applicability to real-time control tasks such as air traffic control, piloting and teleoperation (Vagia et al., 2016). Later on, Parasuraman et al. (2000) began to emphasise that different aspects of human–computer interaction could be automated. The authors proposed that four cognitive functions could be used as input functions, of which, automation could be applied independently. These input functions were defined as information acquisition (the task of sensing, recognising and monitoring information), information analysis (the task of processing, predicting and analysing), decision selection (the task of action selection between different alternatives) and action implementation (the task of responding). While they did not identify levels of automation in the same manner as former taxonomies, they did suggest that automation within each function could range between 'low' and 'high'.

What all taxonomies have in a common is a consensual view that the human operator is expected to carry out all tasks at lower ends of the automation taxonomy, while at the higher end of the taxonomy, automated systems can take on the majority of these tasks. Other transportation domains (e.g. automated metro lines) have typically used reduced representations of the taxonomies outlined above. For example, Georgescu (2006) proposed three operational models for automated metro lines that share similarities with Endsley and Kaber's (1999) definitions. These were semiautomatic (reflecting shared control), driverless operation (reflecting supervisory control) and unattended operation (reflecting full automation). For complex task environments, Endsley and Kaber (1999) suggest that the level of automation can vary between manual control, supervisory control and full automation. With this in mind, it is possible that the same automation pathway could be applied to driving. However, for driving, the jump from manual to supervisory control is likely to be too rapid. This is because drivers need time to understand the effects of the automated systems and how they behave (Stanton et al., 2007a). Rather, the level of driving automation may vary somewhere between manual control (Level 1), decision support (Level 4), automated decision-making (Level 8), supervisory control (Level 9) and full automation (Level 10) using Endsley and Kaber's (1999) definitions. The addition of decision support gives the driver an opportunity to develop their awareness of system state. It is thus analogous to a 'safety gantry', ensuring that the driver builds an awareness of system capabilities and limits.

However, the taxonomies described above are rarely used within industrial practice. Instead, more emphasis is placed upon the National Highway Traffic Safety Administration (NHTSA, 2013), Bundesanstalt für Straßenwesen (BASt Expert

Group; Gasser and Westhoff, 2012) and increasingly upon the Society of Automotive Engineers (SAE) International Standard J3016 definitions of automation, which was recently updated in September 2016. These taxonomies have all sought to better define the functional limits of automated subsystems operating at varying levels of autonomy with specific application to the automotive industry. This in turn has provided some, albeit limited, insight into the *new* role of the driver. This is because these taxonomies have intentionally avoided the use of normative terminologies to avoid the derivation of design requirements. Their interchangeable use within the literature has also led to confusion over what the driver can and cannot do under different levels of automation as while the taxonomies remain 'fairly consistent', they are not identical. Where both NHTSA and BASt taxonomies each define five levels of automation, NHTSA relies upon a numeric system ranging from from Level 0 ('No Automation') to Level 4 ('Full Self-Driving') while the BASt Expert Group uses labels ranging from 'Driver Only' to 'Full Automation'. SAE, in contrast, identifies six levels of automation ranging from zero ('No Automation') to five ('Full Automation'). There is of course a general consensus that as the level of automation increases, the driver becomes increasingly removed from the physical and cognitive tasks of driving. This is reflected in Table 2.1 that shows how some of the primary tasks of driving (Smith, 2013) are allocated between the driver and automated subsystems. Importantly, intrinsic details relating to the new role of the driver is still not properly acknowledged or understood.

Even so, it is important to acknowledge that vehicles operating at SAE Level 4 ('Full Automation') will bring with them new tasks and responsibilities for drivers to engage in. For example, coupling and decoupling from the control-feedback loops remain important tasks (Nowakowski et al., 2015). The entire spectrum of driver responsibilities and workload should therefore be considered in order to deliver an effective 'handover' between the driver and the autonomous vehicle.

THE CHANGING ROLE OF THE DRIVER

There are many lessons that can be taken from the field of aviation as we increasingly see the role of the driver becoming analogous to the role of a pilot (Stanton and Marsden, 1996). In aviation, Hutchins (1995a) described two roles in which the pilot can serve: Pilot Flying (PF) and Pilot Not Flying (PNF). While the PF is responsible for overall control of the plane, the PNF is responsible for communicating with Air Traffic Control and aircraft systems as well as completing all of the checklists that are required during each phase of flight. Thus, the burden of responsibility simply changes rather than being reduced. In recognition of the changing responsibilities of the PNF, the Federal Aviation Administration (FAA, 2003) altered the terminology of PNF to that of Pilot Monitor (PM). In the same way, a Driver Driving (DD) is responsible for overall control of the vehicle while a Driver Monitor (DM) would assume a similar role to that of PM and monitor the behaviour of the vehicle and automated subsystems to ensure safe and normal driving practice is maintained as automated features become engaged. Of course, the introduction of automation into the driving task does not necessarily mean that the driver will assume the role of DM. We already know that the perception of increased reliability can lead drivers

TABLE 2.1

Allocation of Function between the Driver and Automated Subsystems across NHTSA, BASt and SAE Automation Taxonomies

Level of Automation (SAE)	Longitudinal and Lateral Control	Monitoring of the Environment	Operational and Tactical Tasks	Strategic Tasks	BASt Level	NHTSA Level	Endsley and Kaber (1999)	Types of Features
0 – No automation	D	D	D	D	Driver only	0	1	Warnings, for example, blind spot information systems
1 – Driver assistance	D/A	D	D	D	Driver assistance	1	4	Adaptive cruise control
2 – Partial automation	A	D	D	D/A	Partially automated	2		Integrated cruise assist
3 – Conditional automation	A	A	D	D/A	Highly automated	3	7	Highway pilot
4 – High automation	A	A	A	D/A	Fully automated	3/4	9	
5 – Full automation	A	A	A	A			10	Google self-driving car

vulnerable to becoming complacent and over-reliant on automated functionality (Lee and See, 2004; Parasuraman et al., 1993). Without active vehicle control, a DM, for example, could become vulnerable to the onset of boredom or fatigue (e.g. Heikoop et al., 2016; Stanton et al., 1997; Young and Stanton, 2002). Thus, DMs may unintentionally drift in and out of the Driver Not Driving (DND) role. This becomes particularly problematic in instances whereby the automated systems are not able to adequately resolve a scenario without human intervention. If the driver *should* be in the role of DM (i.e. during partial automation) but is in fact behaving in a manner more akin to the role of DND, active control of the vehicle may be transferred to a DND (rather than a DM who is prepared to resume the role of DD) who may fail to respond appropriately due to either a sudden increase in driver workload, reduced SA (Dozza, 2012; Stanton et al., 2011) or as a result of startle (Sarter et al., 1997). Mode transitions between the different driver states or roles are represented in Figure 2.1.

Importantly, the behaviour of a DND is difficult to predict and understand because drivers are free to participate in any task of their choosing. While some may argue that at SAE Level 4 autonomy, the driver's role becomes redundant given the capabilities of the system to a 'minimal risk condition' (Gasser, 2014), this is not strictly the case. The Department for Transport's report (2015) recognises that some SAE Level 4 vehicles may still offer a full set of controls that enable manual driving. This means that at some point the DND could regain control of the vehicle whether this be due to a 'forced' transfer of control due to some form of mechanical failure (e.g. sensor failure), through choice (e.g. the driver may want to abort or change the destination of travel or they may simply want to be in control) or simply because autonomous driving features only operate in *some* driving modes at SAE Level 4. It is therefore important that the DM role is recognised because the DND will need to adopt the role of DM during the exchange of control between them and

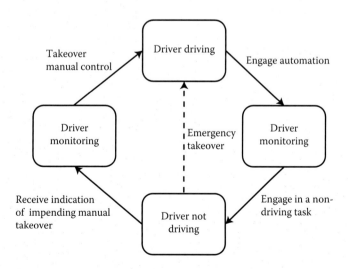

FIGURE 2.1 Driver mode transition network. (Adapted from Banks, V A and N A Stanton. in press. Analysis of driver roles: Modelling the changing role of the driver in automated driving systems using EAST. *Theoretical Issues in Ergonomics Science*.)

the autonomous vehicle. The success of this transition of control will be based upon a number of interacting psychological constructs including SA, workload, trust and skill (Heikoop et al., 2016; Stanton and Young, 2005). If vehicle manufacturers are to handle this transition effectively, a greater understanding of how drivers appraise and make use of higher level autonomy is needed (Richards and Stedmon, 2016). Thus, while less emphasis is placed upon the driver as the level of automation increases, vehicle manufacturers still need to think about ways in which the driver can be supported if and when they choose to regain control of the vehicle, especially during early versions of highly automated driving systems. This is because the DD, DM and DND are closely related and likely to be adopted interchangeably throughout the duration of a drive, especially during the intermediate phases of automation.

While the DND role represents the aspiration of many Original Equipment Manufacturer's (OEMs), there are no such systems that exist on the market today that allow this to happen. If a driver does find themselves in the role of DND, they should be supported back into the role DM to ensure that the overall goal of the system network can be appropriately maintained. For current systems of today, the role of DND could be seen as a form of automation misuse (Parasuraman and Riley, 1997) given the functional limits of automated architecture. Strategies to support the role and maintenance of the DM role are therefore important avenues for further research.

HUMAN FACTORS CONSIDERATIONS IN USING AUTOMATION

To better understand the functionality of the driver–vehicle interactions in using automation, research has typically focussed upon the operationalisation of the following concepts (Saffarian et al., 2012; Sheridan et al., 2008).

REDUCED SITUATION AWARENESS

SA is a multidimensional concept that can describe how individuals (Endsley, 1995), teams (Salas et al., 1995) and systems (Stanton et al., 2006) both develop and maintain their awareness during task performance. Endsley (2006) formerly described SA as

> the perception of the elements in the environment within a volume of time and space, the comprehension of their meaning, and the projection of their status in the near future. (p. 529)

In a dynamic driving environment, individual SA on behalf of the driver is built through the monitoring of critical variables such as speed, road positioning, behaviour of own and other vehicles as well as weather conditions (Walker et al., 2008). SA explains how drivers use this information to combine their long-term goals (e.g. navigation) with short-term goals (e.g. avoiding collisions with other road users) in real time (Walker et al., 2008) while predicting how these variables will continue to change in line with the environment (Gugerty, 1997). Driver SA therefore can be seen as activated knowledge (Salmon et al., 2012) – knowledge that relates to the driving task, at a specific time, within the road environment.

Essentially this means that there can sometimes be a failure to notice change in the external environment due to some form of distraction or interaction with other in-vehicle devices. While it is hoped that vehicle automation will have the desirable benefit of improving the monitoring ability of drivers as they will have more attention devoted to the task at hand, the literature is littered with instances where automation has proven to be problematic as an individual's ability to monitor the visual scene efficiently may actually decrease under automated driving conditions since automation leads to changes in levels of vigilance and complacency (Kaber and Endsley, 2004). For example, on 12 February 2009, a Colgan Air Flight 3407 crashed near Buffalo, New York. A synopsis by the National Transportation Safety Board (2010) reported that the co-pilot incorrectly programmed information into the on-board computers causing the plane to slow down to an unsafe speed triggering a stall warning. The captain had not noticed that the plane had slowed down indicating a lack of SA and responded incorrectly by repeatedly pulling back on the controls, which overrode two safety systems. The correct procedure would have been to push the control yoke forward. An investigation later concluded that there were no mechanical or structural problems that would have prevented safe flight if the pilot had responded correctly to the original problem. All passengers and crew were killed.

A loss of SA was also implicated in the Air France 447 accident in 2009 killing all people on-board (Salmon et al., 2016). A major investigation by the Bureau d'Enquêtes et-d'Analyses (2012) concluded that the fatal incident was a result of a series of events following the disconnection of autopilot as a result of frozen pitot tubes in adverse weather. This led to the plane stalling and crashing into the Atlantic Ocean. The report scrutinised the aircrews' lack of awareness of situation, inability to establish the correct procedure to follow in such an event and an overall failure to control the aircraft.

While aviation is considered to be one of the safest forms of transport, 'human error' is considered to be one of the principal threats to flight safety according to the Civil Aviation Authority (1998). Interestingly, what many incidents occurring within aviation have in common is that functioning automated safety systems are either overridden or ignored placing perfectly serviceable vehicles in otherwise dangerous situations (Stanton and Salmon, 2009). There are two primary forms of error that have been associated with automation in aviation (Stanton and Marsden, 1996). These are mode error (Endsley and Kiris, 1995) and automation surprises (Sarter et al., 1997). Mode errors on modern flight decks started being reported in the late 1980s (Sarter, 2008). A mode error in aviation is typically a result of mode confusion and is characterised by the pilot performing an action that is appropriate for the assumed state rather than the actual state of the system (Mumaw et al., 2001). Numerous studies have shown that pilots can become confused about both system state and the behaviour of flight deck automation (e.g. Sarter and Woods, 1995). Automation surprises are closely linked with pilot mode errors. This is because the pilot perceives that the automation is engaged in activities that were not commanded or engaged (Mumaw et al., 2001). In this way, an automation surprise on the flight deck represents a miscommunication between the automation and human operators leading to a gap in the pilots' understanding of what the system is doing/going to do (Sarter and Woods, 1995; Sarter et al., 1997). These forms of error are not, however, limited to the aviation domain and it seems highly likely that similar errors will occur within driving automation.

In fact, the occurrence of both mode confusion and error in driving with automation is already documented (e.g. Stanton et al., 2011). For example, Andre and Degani (1997) discuss a common error that drivers make when using cruise control. In their paper, they discuss a situation whereby a driver overrides cruise control and increases their speed (in this case slow moving traffic as a consequence of poor weather increased speed once the weather had improved). When they chose to exit the highway, the driver had forgotten that cruise control was still active (thus meaning that the driver experienced mode confusion) and once the vehicle had fallen below the initial set speed, a sudden 'jolt' of acceleration had led them to lose control of the vehicle. Similar errors may also occur when using Adaptive Cruise Control whereby a driver may follow behind a slow moving vehicle (i.e. travelling below their desired set speed) when wanting to exit the highway. As soon as the driver moves the car onto the exit road, the vehicle will increase its speed to reach the preset desired speed. It is at this point, however, that drivers will want to slow their vehicle for the upcoming junction or intersection. Thus, performance impairment may be attributed to a failure to recognise and match external environmental demands.

ERRATIC CHANGES TO DRIVER MENTAL WORKLOAD

Mental workload can be described as the relation between the attentional resources demanded by the task and the resources actually available to complete the task (Sheridan et al., 2008; Singleton, 1989), which echoes the philosophy underpinning Resource Theory. Much of the research available on mental workload comes from the field of aviation and it has provided some worrisome results with many pilots reporting that the use of automated systems actually increases mental workload when it is needed the most (Bainbridge, 1983; Sheridan et al., 2008). In driving, Reinartz and Gruppe (1993) suggested that automation may simply shift driver attention to other tasks such as system monitoring resulting in little reduction in workload.

An alternative perspective comes from Malleable Attentional Resource Theory (MART; Young and Stanton, 2002). MART proposes that there are separate attentional pools, but far from having a fixed capacity, these attentional pools remain robust and are capable of adapting depending on task circumstances. This means that task demand can affect the size of the attentional resource available to complete a task and therefore it can be possible to both overload and underload human controllers. It is possible that high levels of driving automation may lead drivers to become cognitively underloaded (Young and Stanton, 2004), especially in routine situations (Ma and Kaber, 2005; Stanton and Young, 2005). This may resonate as 'highway hypnosis', a form of drowsiness or fatigue that can lower driver alertness (Wertheim, 1978). This altered state may lead to drivers who are unable to respond to changes within their environment in the same way, implicating the concept of SA. The general consensus is that mental workload optimisation is crucial to maintaining effective task performance (e.g. Wilson and Rajan, 1995). Such optimisation inevitably involves a balancing act between demands and resources of both task and operator. However, optimising systems performance during transitional automation would require the driver to remain an active, rather than passive, supervisor of the system (i.e. DM role). Strategies to maintain the driver in-the-loop are therefore an important area of investigation because overall driver workload has been shown to reduce as both the physical and cognitive tasks associated with driving become automated. This workload shift sees the driver transitioning from an active operator (i.e. DD role) to more of a passive monitor (i.e. DND role), which conflicts with the desire to optimise system performance (Kaber and Endsley, 2004).

TRUST, OVER-RELIANCE AND COMPLACENCY

The concept of trust surrounding vehicle automation has largely derived from the idea of complacency (e.g. Lee and See, 2004; Parasuraman et al., 1993; Young and Stanton, 2002). If automation is perceived by the driver to be highly reliable, the driver may not monitor the system as closely as perhaps is warranted and therefore may not expect occasional failures. Thus, the perception of increased reliability instils trust and drivers may become complacent. de Waard et al. (1999) reported that 50% of drivers failed to regain control following system malfunction in a driving simulator study on an automated highway system due to the belief that the system would intervene despite the system being compromised. Although the reality of system failure is small in most cases due to an extensive testing phase, operational failings such as the inability to automate all aspects of the driving task leave the driver vulnerable to the need of intervention whether it is prompted by the system or not (Larsson, 2012).

Over-reliance on automation has been highlighted as a possible cause for several aviation accidents. For example, the National Transportation Safety Board (1994) determined that a pilot who demonstrated low confidence in their manual flying ability was too reliant on automation and failed to monitor aircraft speed during a final approach in a snowstorm causing them to crash land short of the runway near Columbus, Ohio. The Air Transport Administration (1989) and the FAA (1990) have both expressed concern over pilots' reluctance to regain control from automated systems. Worryingly, Riley (1994) found that while novice pilots would turn off automation when it failed, almost half of the experienced pilots did not.

With trust and acceptance emerging as key concepts within the automobile industry, there is a growing body of literature into this area as manufacturers strive to design automated systems that will be widely accepted and adopted.

SKILL DEGRADATION

Skill development and maintenance remains a lasting concern within the field of driving automation, especially if drivers become 'hands and feet free'. This is because automation may hinder the learning potential of future drivers or lead to a loss of skill due to lack of manual input (Lavie and Meyer, 2010; Miller and Parasuraman, 2007; Parasuraman, 2000). Reaching an appropriate level of automation is therefore extremely important to ensure that drivers can maintain an appropriate level of driving experience (Patten et al., 2006). However, a number of studies have indicated that performance under increased levels of automation can decline, suggesting that automation can negatively impact driving performance through skill degradation. For example, Jameson (2003) reported that individuals perform manual tasks less efficiently once they have acclimatised to the system performing the task on their behalf.

However, this may only be a problem in the short term while increasing levels of automation are introduced into the driving system. Once fully automated vehicles are fully integrated into our transportation network, issues relating to skill degradation may become less problematic. Even so, while we remain in the intermediate phases of automation, skill degradation is a real concern.

CONCLUSIONS

Taking a systems view, it seems that the mediation of activity may be better coordinated if an augmented approach is taken. The design of technology that can integrate our own cognitive ability and can be used to further extend our capabilities is achievable if the interactions between individuals, environment and other media are explored.

It is clear to see that regardless of the automation framework that is adopted (e.g. theoretical or practical), the terminologies used within them remain fairly similar. It is essential that agreement can be reached on the appropriate 'label' assigned to automated systems and that its usage remains consistent throughout the design process to avoid confusion about system limitations and functional boundaries. Even so, it seems that at least for the interim period between transitional automation and full vehicle automation, control transitions will continue to be made between the driver and the automated system due to issues surrounding practicability, liability and individual preferences of the driver (SMART, 2010). For this reason, the driving task can be best described as shifting somewhere on a continuum between manual and fully automated driving. Thus, rather than being either strictly manual or automated, the driving task is shared between the driver and the automated system. In essence, what this means is that the whole driving system can involve the automation of different processes at different levels simultaneously. Until drivers become completely 'hands and feet free', serious concerns remain with regard to out-of-the-loop performance problems (Billings, 1988; Endsley and Kaber, 1997; Endsley and Kiris, 1995) and the ability of the driver to detect and resolve errors in their new supervisory role (Endsley and Kiris, 1995).

FUTURE DIRECTIONS

The allocation of system function between the driver and automated subsystems within an automated driving system appears to be key in understanding how the role of the driver will be affected by automation implementation. However, our understanding of network dynamism is limited to the definitions outlined by SAE, BASt and NHTSA who fail to appropriately capture the changes occurring within the control-feedback loops of driving at varying levels of autonomy.

While automation taxonomies go some way in describing the workload shift between the driver and the automation, more explicit modelling is required to truly understand how the driver's role within the driver–vehicle control loops is impacted by automation implementation. Chapter 3 builds upon Chapter 2 by introducing the concept of Distributed Cognition (Hutchins, 1995a,b). This approach recognises that interactions can occur between human and non-human agents. Modelling the communication patterns that exist between these agents will further extend our understanding of (a) how these systems function and (b) the relationships that exist between multiple system agents at varying levels of automation.

3 Adopting a Systems View in the Design of Automated Driving Features

INTRODUCTION

Systems Engineering can be seen as an interdisciplinary approach to the field of engineering. It integrates both technical and human-centred approaches to look more closely at work processes, optimisation and risk management. This holistic approach is concerned with how the functioning and performance of a joint cognitive system can be best described and further understood. The driving task is an example of a joint cognitive system (Salmon et al., 2008), one that comprises the driver and the devices in which they engage. This viewpoint stems from the belief that every 'agent' within a system plays a critical role in the successful completion of a task, and more importantly, 'agents' can be both human and non-human (Salmon et al., 2008; Stanton et al., 2006). It therefore provides an analytical framework that can be used in the design of adaptive automation (Hollnagel and Woods, 1983). Artman and Garbis (1998) suggested that cognition is achieved through close coordination of the elements or agents involved in the system, and in a vehicle, both the driver and in-vehicle devices are seen as 'agents'. It appears to be team cognition that is the binding mechanism that produces coordinated behaviour (Cuevas et al., 2007). Although early research into automation seemed to focus heavily upon autonomy (i.e. full automation), current research now focusses upon satisfying the requirements of joint activity, including human–machine teamwork (Klein et al., 2004). Understanding system functioning however is extremely complex because the 'behaviour' or interaction that occurs between system components is not always well defined or understood. The aim of this chapter is to better define and characterise subsystem behaviour within automated driving systems using Distributed Cognition to explore whether or not automated subsystems fundamentally change the driving task by affecting the ways in which the driver interacts with vehicle systems.

DISTRIBUTED COGNITION ON THE ROAD

Unlike traditional theories, Distributed Cognition goes beyond the individual and encompasses the interactions that take place between humans, resources and materials within their environment across space and time (Hollan et al., 2000; Hutchins, 1995a). It therefore fits nicely within the Systems Design paradigm by recognising that both human and non-human agents are vital to the flow of information within a system (Griffin et al., 2010). With vast amounts of information exchange between multiple agents, the ability to sense changes within different representational states, understand them and then perform some form of computation to deal with these changes implicates SA (Endsley, 1995) and describes the essence of Distributed Cognition. Although the study of SA originated within the aviation domain (Stanton et al., 2001), Endsley (1995) identified SA as a critical component in driving.

Distributed Cognition has been most famously applied to description of task partitioning in a pilot's cockpit (Sorensen et al., 2011), and there appears to be no reason why it cannot be applied to driving. From this perspective, SA is formulated through a myriad of individual components and cannot be predicted based solely upon one of these individual components or the mere combination of individual SA from different agents (Salas et al., 1995). This idea is particularly relevant to vehicle automation because the driver uses assistive aids to help build a 'picture' of what is happening in the world (Walker et al., 2010). There is a need to move away from traditional notions of SA (Endsley, 1995) that dominate Ergonomics at present to one that focuses upon entire systems (Gorman et al., 2006; Salmon et al., 2008; Sorensen et al., 2011; Walker et al., 2010). This is because there are very few complex tasks that can be performed on a completely individualistic basis (Perry, 2003; Walker et al., 2010). Distributed Situation Awareness (DSA; Stanton et al., 2006, 2015) offers a compatible approach that assumes SA is a system-level phenomenon rather than individual-orientated (Salmon et al., 2008; Stanton et al., 2006). DSA outlines that SA can be held by human and non-human agents, that different agents view their environment differently and that at an individual level SA overlap will be dependent upon the goals of each agent. DSA also recognises that communication can be non-verbal and that SA loosely holds systems together whereby one agent has the ability to compensate for degraded SA in another (Stanton et al., 2006, 2015).

There is a clear need then to consider the interactions that occur between multiple agents as being 'cooperative' because automated systems can become 'vital non-human' agents within the task under the right circumstances (e.g. Cuevas et al., 2007, p. B64). According to Cuevas et al. (2007), a human–automation team can be defined as the coupling of both human and automated systems that must work both collaboratively and in coordination to successfully complete a task. It is important that the principle of complementarity is adopted, with the allocation of tasks serving to maintain control while retaining human skill (Grote et al., 1995). As with Free Flight (Langan-Fox et al., 2009), driving automation poses many challenges with regards to the interaction between humans and automation including operational functionality and system management. There may be confusion over who (the driver or automated subsystem) has authority over 'which' vehicular controls as the level of automation increases.

Distributed Cognition in driving provides a means to employ Human Factors insights into the early phases of the design process (Jenkins et al., 2009; Walker et al., 2015). It aims to provide a clearer understanding of task partitioning between the driver and automated subsystems and recognises that the cognitive processes normally completed by the driver can be shared across this system (Hollnagel, 2001; Stanton, 2014a,b) to achieve a common goal (Hoc et al., 2009). Up until now, it is an approach that has been successfully applied to a number of domains including ship navigation (Hutchins, 1995b), airline cockpits (Hutchins and Klausen, 1996), engineering practice (Rogers, 1993), search and rescue (Plant and Stanton, 2016) and air traffic control (Halverson, 1995). The application of Distributed Cognition to driving is a new and unexplored medium, yet there appears to be great benefit in doing so in terms of automation development and system safety.

SYSTEMS DESIGN FRAMEWORK

The Distributed Cognition approach makes use of a number of exploratory methods including detailed analysis of real-life events, network simulations and laboratory experiments (Rogers, 1997). The authors of this book propose that Distributed Cognition can be applied to driving using the combination of traditional task analysis and qualitative research methods following a two-phase Systems Design Framework (see Figure 3.1). It represents a step-by-step framework that provides a basis for future exploration into the design and allocation of system function for both pre-existing technologies and the development of future automated systems. Split into two phases, the Systems Design Framework provides researchers with the tools and techniques required to model system behaviour (Phase 1) as well as outlined the processes that can be used to validate the assumptions made within these models and identify areas of weakness in order to improve overall systems design (Phase 2).

Phase 1 (Modelling)

Step 1: Design Idea, Concept or Prototype
The first step of applying Distributed Cognition to driving involves determining which specific driving task the investigator is interested in. With this in mind, it is possible to modify and extend current vehicle technologies and/or develop future automated driving systems. Focus group discussions or mind-map exercises are a useful starting point. This enables the investigator to identify the system components, or agents, required to complete the task as well as to outline functional boundaries.

Step 2: Allocation of Function
The allocation of system function should be viewed as a high-level task analysis that aims to give a general impression of the workload shift between system agents from

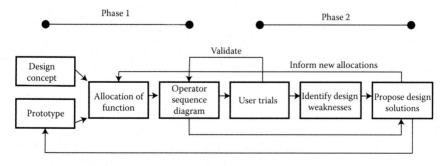

FIGURE 3.1   Systems Design Framework divided into two phases.

manual to fully automated control in a descriptive manner. This enables researchers, systems designers or engineers to further explore and discuss allocation of system function. With the driving task consisting of so many subtasks (see Walker et al., 2001, 2015), automation can be applied to different aspects of 'driving' at differing levels of autonomy (Endsley and Kaber, 1999). Once automated subsystems are activated, the driver assumes that a degree of control has been assigned to the system. This can change the nature of driver–vehicle interaction as a division of labour occurs. Echoing the viewpoint of Endsley and Kaber (1997), it must be ensured that the driver knows exactly 'who' is expected to do 'what' during the driving task to maintain safe operation. Thus, in order to design a human-centred product, the designer should augment the task in which the product is designed for. Much like Hutchins (1995a) described two roles within a pilot's cockpit (Pilot Flying and Pilot Not Flying), there are two primary *actors* (note, not roles) capable of controlling a vehicle (a driver and an automated system).

A mapping exercise such as that presented in Table 3.1 can achieve the desired output. In this example, a matrix was developed to show how the completion of individual information-processing functions may be shared between the driver and an automated system (Banks et al., 2014a). Importantly, the allocations assumed within Table 3.1 are intended to be descriptive rather than normative but offer a useful starting point in thinking about how allocation of function may be shared between the driver and automated systems. In addition to the information-processing functions outlined in previous taxonomies (e.g. Endsley and Kaber, 1999; Parasuraman et al., 2000). Banks et al. (2014a) propose two additional functions to make them more applicable to driving. Firstly, an anticipatory phase took into consideration feed-forward mechanisms occurring within the driving task. For the driver, much of this will be based upon prior experience, but feed-forward information can also come from other system agents. Secondly, a recognition phase took into account the distinction between object detection and object recognition within the driving task. For example, Banks et al. (2014a) propose that the identification of a pedestrian stepping out or a car joining the carriageway can be recognised only if they are first detected. In the recognition phase, a greater degree of attention must be devoted to an object of interest. This means that potential hazards within the visual scene will be allocated with additional attentional resources so that they can be identified and either confirmed or discarded. This pattern of behaviour is important for later decision-making and strategy generation processes. What this means is that drivers monitor and acquire information relating to their environment and anticipate likely events and interactions that may occur (Endsley and Kaber, 1999; Parasuraman et al., 2000). When an event or interaction does occur, the driver must firstly detect it and recognise the salient features of the event. On the basis of this recognition, the driver must then decide on strategies for dealing with the event (Endsley and Kaber, 1999), select the most appropriate strategy (Endsley and Kaber, 1999; Parasuraman et al., 2000) and perform a response to implement this strategy (Endsley and Kaber, 1999; Parasuraman et al., 2000).

For our readers' clarification, the following descriptions provide narrative examples of how automated systems use information within the environment to support decision-making and response execution.

TABLE 3.1

Distributed Cognition of Information-Processing Functions within the Driving System When Systems Are Active (Intended to Be Descriptive Rather Than Normative)

Technology	Distributed Cognition of Information Processing Functions						
	Monitor	Anticipate	Detect	Recognise	Decide	Select	Respond
Parking Aid (Sensors/ Beeps)	D/A	D	D/A	D	D	D	D
Power Steering	D	D	D	D	D	D	D/A
Night Vision	D/A	D	D	D	D	D	D
LDWS	D/A	D	D/A	D/A	D	D	D
Collision Warning	D/A	D	A	D/A	D	D	D
Blind Spot Information System	A(D)	D(A)	A	D	D	D	D
Pedestrian Detection System	A(D)	D(A)	A	D/A	D/A	D/A	D/A
Automatic Braking	A(D)	D(A)	A	A	A	A	D/A
Intelligent Light System	A(D)	D(A)	A	A	A	A	A
Intermittent Windscreen Wipers	A(D)	D(A)	A	A	A	A	A
Collision Avoidance	A(D)	A	A	A	A	A	A
ABS	A(D)	A	A	A	A	A	A
Traction Control	A(D)	A	A	A	A	A	A
ACC	A(D)	D(A)	A	A	A	A	A
ACC Stop and Go	A(D)	A	A	A	A	A	A
Lane Keep Assist	A(D)	A	A	A	A	A	A
Park Assist	A(D)	A	A	A	A	A	A
Driver Monitoring	A(D)	A	A	A	A	A	A

Note: D, driver; A, automation; D/A, driver and automation can perform function; D(A), driver should ideally perform function but subsequent automation involvement has been designed to overcome driver inactivity in this function; A(D), automation capable of performing function but the driver is still expected to continue active monitoring of both subsystem behaviour and events in the driving environment.

- *Parking Aid*: Benjamin wants to park his car in a parking lot. He drives around the parking lot in search of a space. He notices a gap ahead of him and at closer inspection realises that it is a free space. He looks at the environment around him and decides that it is big enough to park his car safely and that he would also like to reverse into it. Pulling his car forward ahead of the parking space, Benjamin checks his mirrors and selects the appropriate gear. Using his previous experience, Benjamin is able to reverse safely into the gap. However, he is unable to see the back wall clearly due to poor lighting. His Parking Aid begins to 'beep', which must mean that he is getting close to the wall. In this example, the Parking Aid has been monitoring its environment and detected that the distance between the extremities of the car and obstacles within the environment is becoming reduced. It has alerted Benjamin to such obstacles using an auditory signals, of which will become more panicked the closer he gets to the wall.

- *Adaptive Cruise Control (ACC)*: Susan is driving her car along a highway. She engages ACC – a form of intelligent cruise control that can slow down or speed up depending upon the traffic situation ahead. The driver sets a maximum speed which the vehicle is now responsible for maintaining. While Susan is still expected to monitor her surroundings and anticipate the movements of other road users in case she may need to override the system, the ACC system must also monitor the roadway ahead and detect any obstructions in the path of the vehicle. It uses radar sensors to monitor the traffic ahead and can 'lock' on to a vehicle ahead in lane and maintain a 2- to 4-second gap depending on driver preferences. Sometimes, recognition of a vehicle ahead is shown within the HMI. If the lead vehicle slows down, or another vehicle enters the lane ahead, radar headway sensors detect the object and digital signal processors send signals to the engine or braking system to decelerate. This deceleration signal is to preserve the safe gap between vehicles. Once the road is clear, or the vehicle ahead speeds up, the sensors will send signals to reaccelerate to the set speed.

Table 3.1 shows how the driver and automated subsystems can work in parallel or independently to one another in a selection of automated features available within our vehicles. Using this representation, it becomes clear that many of these features operate at an enhanced level of automation as they are capable of performing all seven information-processing functions. However, it is important to note that there appears to be a greater difference in the level of automation between assistive technologies (e.g. Collision Warning Systems) and the level of automation demonstrated by controlling technologies (e.g. Collision Avoidance). Thus, the level of assistance that the driver receives between individual subsystems in these broad categories of 'Assistive' and 'Controlling' technologies varies greatly. For example, a Parking Aid provides assistance in only one function whereas Collision Warning System and Lane Departure Warning System (LDWS) can assist the driver in multiple functions. In contrast, controlling technologies such as ACC and Park Assist have been designed to perform all seven information-processing functions so that a greater degree of control transfer can occur between the driver and automated subsystems. In terms of Distributed Cognition, controlling technologies assume responsibility for all seven processes.

Notably, individual features acting alone (i.e. without driver monitoring) are unlikely to create safer driving environments due to their functional limitations (Stanton et al., 2011). This perhaps explains why, in recent years, it is becoming increasingly prevalent to use multiple automated subsystems simultaneously. In the case of active hazard detection for instance, combined Pedestrian Warning/Detection Systems (Pedestrian AEB), Collision Warning Systems and Active Braking will bring about enhanced levels of automation (SMART, 2010). However, subsystem synergies such as this mean that the level of assistance provided to the driver across the entirety of the driving system becomes increasingly complex (Stanton et al., 2011). This is because the level of automation is technology specific and the driver will need to remain aware of multiple system states simultaneously (Cuevas et al., 2007; Dehais et al., 2012; Walker et al., 2009).

The mapping of Allocation of Function should not be viewed as a mandatory exercise when applying Distributed Cognition to driving. Instead, it should be viewed as an opportunity to discuss with peers what individual system agents may be responsible for during automated driving. It should not be viewed statically because in dynamic environments such as driving, allocation of function is likely to change depending upon context (Hancock and Scallen, 1996).

Step 3: Operator Sequence Diagrams

We have already seen from the matrix in Table 3.1 that as vehicles become increasingly 'intelligent', they become better able to take over elements of the driving task traditionally performed by the driver (Dingus et al., 1997; Walker et al., 2001). The matrix alone does not however reflect the interactions that occur between the driver and vehicle subsystems. To further our understanding of Distributed Cognition in driving, it is vital that these interactions are modelled to ensure that the introduction of automation does not negatively impact upon the functionality of the driving system (Hoc et al., 2009). Being able to analyse and evaluate the activity of system agents within the driving system provides an opportunity to establish how drivers may recover in the event of system failure. This becomes increasingly important as technologies become more intelligent (Shorrock and Straeter, 2006) and we remember that the driver and automated systems become more analogous to a 'team' (Cuevas et al., 2007). Automated features operating at higher levels of autonomy have the potential to play an important role in detection, problem-solving and decision-making functions that would otherwise be completed by the driver under manual conditions. Thus, Step 3 should be viewed as the opportunity to explore a task network with more situational context.

The application of OSDs (Brooks, 1960; Kurke, 1961) provides a novel way to investigate Distributed Cognition within automated driving systems. OSDs provide clear, easy-to-read graphical representations (see pp. 34–37 for examples), and since the 1960s, the method has been widely used in Systems Engineering. They have been applied successfully to a wide range of domains including air traffic control (Walker et al., 2005), rail (Walker et al., 2006), energy distribution (Salmon et al., 2004), nuclear industry (Kirwan and Ainsworth, 1992) and most famously, to a pilots cockpit (Sorensen et al., 2011). It is one of the most cost-effective ways to simulate a complex system (Chapanis, 1995) and can be described as a paper-based methodology that can be easily implemented at different stages of the design process from a design concept through to prototype.

Outputs seek to contribute to our understanding of system dynamism and operation as they provide a means to assess weak links between agents and communication flows (Griffin et al., 2010; Kurke, 1961). According to Wallace et al. (2000), OSDs must adequately capture and represent the subtasks and operations required to complete the task including subevents, decisions, capabilities and the ordinal or temporal flow of control and information by using standard geometric figures. These figures are individually coded to denote different elements of the operational sequence (Kurke, 1961).

Although OSDs can be criticised for being task specific and therefore not capable of representing the true complexity of system functioning, this also serves as a benefit if the analyst is interested in simulating a 'specific' subtask conducted within a complex system. According to the Hierarchical Task Analysis of Driving (Walker et al., 2015), there are approximately 1600 identifiable driving subtasks of varying complexity. This means that within the driving domain, OSD methodology could be used to simulate individual subtasks of driving that make use of additional hardware and automated technology. This makes it possible to identify how the driving task may change as a result of technology introduction (pre- vs. post-automation) and be a useful mechanism in comparing and contrasting different technological formats (e.g. display units). OSDs therefore offer an easy way of visualising the interrelationships and communications that occur between different agents, providing an inexpensive alternative to mock-ups and prototypes that attempt to address the same purpose.

While in their simplest form, OSD representations provide a qualitative overview of how a system may function, it is also possible to interrogate these networks using quantitative network metrics. This is achieved by creating a Matrix of Association – simply counting the number of links between nodes and inputting them into a tabular format. The benefit of subjecting OSD representations to network analysis is that it becomes possible to identify how network dynamism changes as you manipulate or alter elements of the system. Within the literature, the following network metrics have typically been applied to driving systems (Salmon et al., 2009; Walker et al., 2011).

Network Density   Network density represents the level of interconnectivity between nodes. It is expressed as a value between 0 and 1. A score of 0 represents a network that has no connections between nodes whereas a score of 1 represents a fully connected network (Kakimoto et al., 2006). Higher scores indicate an enhanced level of system awareness as there are a greater number of links between nodes (Walker et al., 2011). The formula for network density is presented below (adapted from Walker et al., 2009):

$$\text{Network density} = \frac{2e}{n(n-1)}$$

Network Diameter   Network diameter is used to analyse the connections and paths between nodes within the networks (Walker et al., 2011). Greater diameter scores reflect an increased number of nodes within the network while denser networks have smaller values. Smaller values simply reflect that the route through the network (e.g. from driver monitoring to subsequent response) is shorter and more direct. It is calculated using the following formula (adapted from Walker et al., 2009):

$$\text{Diameter} = \max_{uy} d(n_i, n_j)$$

Network Cohesion   The cohesion of a network represents the number of reciprocal connections divided by the total number of possible connections (Stanton, 2014a).

Sociometric Status   Sociometric status, rather than analysing the entirety of the system network, focusses upon the analysis of individual nodes and gives an indication of node prominence within the system network as a communicator with others in the network (Houghton et al., 2006). It is calculated using the following formula:

$$\text{Sociometric status} = \frac{1}{g-1} \sum_{j=1}^{g} (x_{ji}, x_{ij})$$

where g is the total number of nodes in the network; and i and j are individual nodes and are the edge values from node i to node j (Salmon et al., 2012). Nodes with high sociometric values are highly connected with other nodes within the system network whereas nodes with low sociometric status values are likely to reside on the peripheral edges of the network (Salmon et al., 2012).

Phase 2 (Validation)

Step 4: User Trials

Although task analysis and its associated methods, such as those outlined in the preceding steps, are a popular and widely used method to assist in the design and development of automated technologies (Putkonen and Hyrkkänen, 2007), it remains a challenge to capture both the cognitive and behavioural elements of a task (Patrick, 1992). In terms of driving, it seems reasonable to suggest that the cognitive processes involved in completing the driving task play the most important role in performance, especially when considering that it is cognition that shapes subsequent behaviour. Although Annett (2004) argues that task analysis does involve cognition and behaviour, Shepherd (2000) suggests that we should begin to consider 'how' this can be represented. Allocation of function matrices and OSD representations are not capable of doing this, so to understand the effects of automation on driver cognition, other methodologies such as VPA (Ericsson and Simon, 1993) could be useful extension methodologies to explore this further.

Driver–vehicle interaction can be explored in driving simulator, closed-circuit and on-road settings. There are a number of advantages and disadvantages associated with each of these strategies. Most driver behavioural studies are conducted in driving simulators as they provide the safest environment (de Winter et al., 2012; Kaber et al., 2012; Stanton and Pinto, 2000; Stanton et al., 2001; Vollrath et al., 2011). This is because there is no danger to the driver or other road users. This makes it possible to investigate driver behaviour in critical driving scenarios. There are also other advantages to using driving simulation over other methods including ease of data collection and versatility. Computer systems provide online data processing, storage and automatic arrangement of data. Investigators can choose parameters of interest meaning that they can collect as much or as little data as they see fit (Bella, 2008; Godley et al., 2002; Nilsson, 1993). They can also be easily configured to simulate a variety of research scenarios (Blana, 1996). This makes it possible to evaluate viable system approaches from numerous alternatives before field testing occurs at a relatively low cost. Vehicle characteristics, such as steering ratios and brake calibrations, can be changed quickly allowing for immediate testing. However, while driving simulation can be tightly controlled (i.e. every driver experiences exactly the same testing scenario), the extent to which behaviour corresponds to what would actually happen in real life is questionable (Blana, 1996). This is because the social and economic factors that often influence driving behaviour are absent. Greenberg and Park (1994) suggested that this alters behaviour substantially. Even so, the study of critical scenarios is difficult, if not impossible, to achieve in the real world (Bella, 2008; Blana, 1996; de Winter et al., 2012; Moroney and Lilienthal, 2009; Nilsson, 1993). One other weakness associated with the use of driving simulators is the risk of simulator-induced motion sickness. Oron-Gilad and Ronen (2007) predict 10% of people will experience some level of sickness.

Closed-circuit or test-track studies represent a step closer to real-world driving (NHTSA, 1997). However, the extent to which they represent real driving behaviour depends upon the nature of the study (NHTSA, 1997). The presence of other vehicles on the test track, for example, can introduce the risk of real consequence (e.g. collision) and is likely to encourage realistic driving practice. Even so, test-track studies offer a more controllable environment over on-road studies (Bach et al., 2009). Of course, on-road studies offer the highest level of fidelity and validity but are coupled with high risk, low versatility and low controllability (Bach et al., 2009; NHTSA, 1997).

Step 5: Identify Design Weaknesses

Upon analysis of the data generated through Steps 3 and 4, it becomes possible to highlight potential system design weaknesses that could negatively affect overall system functionality. For example, performance data such as response times and stopping distances may provide a useful insight into how the introduction of an automated feature into the driving system may affect overall system performance (e.g. in terms of driver behaviour and vehicle dynamics). Qualitative research methodologies enable the investigator to explore other issues relating to systems design that would otherwise be missed from data collection. For example, subjective ratings of trust (e.g. Jian et al., 2000), acceptance (e.g. van der Laan et al., 1997) and workload (e.g. Hart and Staveland, 1988) are generally captured as supplementary material in addition to objective performance data. However, the authors argue that in order to properly explore Distributed Cognition and the allocation of system function from the driver's perspective, driver verbalisations become a crucial data source.

Driver verbalisations can be captured in a number of ways. Think-aloud (Ericsson and Simon, 1993), Critical Decision Method (CDM; Klein et al., 1989) and video-cued recall (Erlandsson and Jansson, 2007) all represent techniques that can be used. The subjective accounts afforded by such techniques can then be subjected to thematic analysis. Thematic analysis essentially aims to identify, analyse and report patterns within data (Braun and Clarke, 2006). It generates further insight into how information is processed and influences our observable behaviour. For driving in particular, this means that it is possible to explore a multitude of areas including system usability and decision-making.

Step 6: Propose Design Solutions
Recommendations for alternative strategies can be raised following the appraisal of results if required. Any change to systems design should revisit task augmentation, modelling and user trial strategies to ensure the success of later prototypes.

PHASE 1: AN EVALUATION

In order to demonstrate the utility of the Systems Design Framework outlined in Figure 3.1, a case study of driver–vehicle interaction at increasing levels of autonomy within driving emergencies has been selected for further exploration into the diminishing role of the driver as more physical control is transferred from the driver to automated systems. Notably, only the first phase of the framework was applied at this stage to demonstrate its potential in describing system-level interaction occurring at varying levels of automation. The rationale behind the use of this example is discussed below.

With driving requiring the driver to continually process the information presented to them in the environment (Fuller, 2005), the level of task demand is determined by a number of interacting factors including environmental (visibility, road markings, signals, camber angles, etc.), social (other road users that may occupy critical areas in driver trajectory), operational (vehicle displays, lighting and control) and finally elements that the driver has direct control over such as speed and vehicle trajectory (Fuller, 2005). Access to this flow of information is based upon the availability of attentional resources, driver perception and decision-making processing (Wickens and Holland, 2000) and will vary depending on the distribution of system complexity, rate and element of certainty (Fuller, 2005). Abnormal or atypical driving upsets this flow of information as it brings additional complexity into the driving task. This is because the driver is required to choose an appropriate strategy to cope with an otherwise 'uncertain' event that they may not have experience in dealing with. One way of coping with this additional complexity is to design automated systems that are capable of improving the safety of both the driver and other road users in addition to traditional active safety technologies such as traction control that are always active.

The following sections are divided into the first three steps of the Systems Design Framework and discussed in turn.

STEP 1: IDENTIFICATION OF DESIGN CONCEPT

There has been growing concern surrounding the safety of vulnerable road users, particularly adults, over recent years (Parliamentary Advisory Council for Transport Safety commissioned report – Road Safety Analysis, 2013). Formerly, vulnerable road users are defined as

> non-motorised road users, such as pedestrians and cyclists as well as motor-cyclists and persons with disabilities or reduced mobility and orientation
>
> **Intelligent Transport Systems Directive, 2014**

This may be attributed to the fact that vulnerable road users are in closer proximity to other road vehicles in urban environments and have significantly less protection.

Although there are a number of mitigation strategies available to local authorities in urbanised developments such as dedicated cycle lanes, pedestrian zones and 20 mph zones, road safety may also be improved through vehicle design (e.g. World Health Organisation, 2004).

It seems likely that vulnerable road users are likely to benefit indirectly by some form of 'intelligent' or 'adaptive' automation designed to promote road safety. Adaptive Headlights, Blind Spot Monitoring and Pedestrian Warning with AEB (Pedestrian AEB) are just some of the technologies that Euro NCAP are introducing more stringent testing procedures for. Even so, there is still debate over whether the use of such systems fundamentally changes the driving task by affecting the ways in which the driver interacts with vehicle subsystems (Banks et al., 2014a). Much of the available research into active pedestrian safety focusses on the technical limits of automation rather than addressing Human Factors in systems design (e.g. Gandhi and Trivedi, 2007; Keller et al., 2011; Rosen et al., 2010). This means that the changing role of the driver is not being recognised. It is clear that there is a need to balance the indirect benefits of automated technology on vulnerable road users with the potential performance impacts on the driver. If, for instance, automation simply removes the need for the driver to monitor their environment, vulnerable road users may benefit from technology fitment but the driver may not (see Stanton and Pinto, 2000, for a review on the possible ill-effects of risk compensation).

Pedestrian AEB was identified by the Euro NCAP as a critical safety system to be widely deployed from 2016 onwards. From a Human Factors and Ergonomics point of view, this gives an opportunity to investigate the impacts of Pedestrian AEB on driver behaviour prior to mass production. Although an extensive amount of literature has been produced surrounding the safety and efficiency of AEB (e.g. Grover et al., 2008), very little research has emphasised Human Factors in its design. It is unclear then how driver decision-making may be affected by the addition of this form of system.

Pedestrian AEB uses radar- and camera-based technology to monitor and detect objects that enter the path of the vehicle and warn the driver about potential collisions. A functional diagram of this system is presented in Figure 3.2. Using pattern recognition and classification within image processing, the system can track pedestrian movements. If the system calculates that the risk of collision is high, it responds by initiating emergency braking and in some instances provides a warning to the driver (Gandhi and Trivedi, 2007).

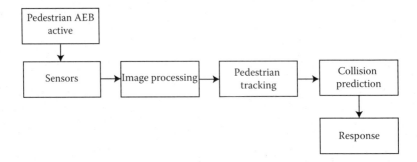

FIGURE 3.2 Functional diagram of pedestrian AEB.

TABLE 3.2

Descriptive Representation of Distributed Cognition of Information-Processing Functions Involved in Pedestrian Detection

Level of Auto	Distributed Cognition of Information Processing Functions						
	Monitor	Anticipate	Detect	Recognise	Decide	Select	Respond
Manual	D	D	D	D	D	D	D
Decision support	D	D	D/A	D	D	D	D/A
Automated decision-making	D/A	D	A	A	A	A	A
Full automation	A(D)	A	A	A	A	A	A

Note: D, driver; A, automation; D/A, driver and automation capable of completing function; A(D), automation capable of completing function but driver still expected to take part.

Step 2: Allocation of Function

As a starting point, Table 3.2 offers useful insight into how increasing the level of automation of pedestrian detection may change the dynamism of driver–automation interactions. It enables us to see how workload begins to shift to the automated system. The levels of automation chosen here represent the hypothetical automation pathway that was identified using Endsley and Kaber's (1999) definitions in Chapter 2. In manual driving scenarios, the driver is responsible for completing all of the physical and cognitive work load associated with the driving task. As the level of automation begins to increase, the automated system is able to assist and eventually control different aspects of the driving task. For example, an auditory warning (reflecting decision support) can assist the driver in detecting critical pedestrian events and facilitate or provide a response on behalf of the driver (e.g. braking support). This mapping exercise is however unable to relay the complex interaction that takes place between system agents to the reader. For this reason, we need a simple, easy-to-read graphical representation of Distributed Cognition so that the reader can quickly see how the level of automation impacts upon the dynamics of the driving system.

Step 3: OSDs for Pedestrian Detection

To explore the utility of OSDs for describing driver behaviour in an automated driving system, four OSDs were developed to represent the interaction that may occur at each level of automation as described in Table 3.2. These visual representations show the workload shift more succinctly and make use of standardised geometric features. Table 3.3 reports the meaning of each of these features.

The OSD presented in Figure 3.3 represents Level 1 automation (manual control) and shows that the driver is responsible for completing all of the physical and cognitive tasks associated with driving. In this way, avoiding a collision with vulnerable road users is based upon driver attentiveness and the ability to adapt to the

TABLE 3.3

Meaning of Geometric Shapes Used within OSD Representations

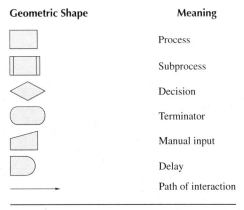

Geometric Shape	Meaning
	Process
	Subprocess
	Decision
	Terminator
	Manual input
	Delay
	Path of interaction

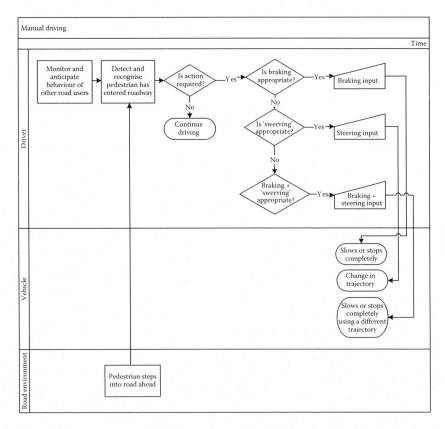

FIGURE 3.3 Manual control of pedestrian detection (Level 1) assuming that the driver is alert and motivated.

ever-changing road environment in the visual scene. In an ideal situation, a collision can be avoided if the driver is able to anticipate vulnerable road users coming into the path of the oncoming vehicle and to brake and/or steer away from them. Alternatively, if the driver decides that the pedestrian or vulnerable road user will move out of the vehicle path before reaching their location, they may choose to take no action.

Figure 3.4 represents Level 4 automation (decision support) and shows that the driver has principal control over the decision-making and selection function but the automation is assisting the driver in the detection phase of information processing by providing an auditory and sometimes visual warning when a threat is identified. The use of an auditory signal allows drivers to become aware of a potential hazard in the vehicle path if they have not yet recognised a threat, which signals the development of DSA (Stanton et al., 2006, 2015). This is important because the driver and Pedestrian AEB collaboratively form DSA. Even so, it seems unlikely that driver workload will be reduced as the OSD suggests that the driver completes the same processing as before, although the complexity of the task with the addition of a warning appears to increase. This is because the addition of a warning is not capable of

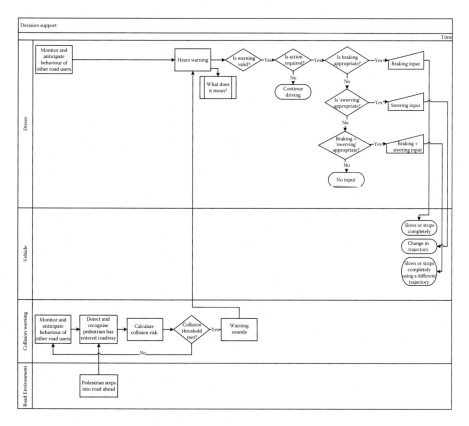

FIGURE 3.4 Decision support in pedestrian detection (Level 4) assuming that the driver has failed to detect developing hazards.

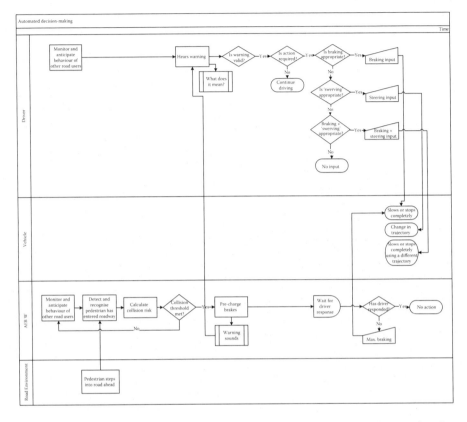

FIGURE 3.5 Automated decision-making of pedestrian detection (Level 8) assuming that the driver has failed to detect developing hazards. Assumes AEB is coupled with an auditory warning.

telling the driver where the 'threat' is located. It may however increase the efficiency of the search and alert the driver to task-relevant information.

Figure 3.5 represents Level 8 automation (automated decision-making) and shows that in addition to an auditory warning, Pedestrian AEB systems also incorporate an element of collision mitigation. At this point, there is some overlap between Level 4 and Level 8 automation. This is because if the driver fails to intervene and a collision is considered to be imminent, Pedestrian AEB is capable of performing an emergency stop, thus automating the decision-making process. This also confirms that the 'automation pathway' is not a stepwise process. A combination of Endsley and Kaber's (1999) definition of automated decision-making and supervisory control sees Level 8 automation capable of generating, selecting and implementing strategies. For pedestrian detection, the nature of the interactions outlined in Figure 3.5 shows that workload is weighted more heavily towards Pedestrian AEB rather than the driver. At Level 8 automation, the main role of Pedestrian AEB is to monitor the behaviour of the driver and intervene when collision risk thresholds have been reached. The main priority is to mitigate the effects of collision.

Although Pedestrian AEB as a single system is not a candidate for full automation, it seems likely that a complex synergy of multiple subsystems may incorporate an element of hazard detection. Assuming that the blend of subsystems allows for the automation of braking and steering inputs, the OSD presented in Figure 3.6 may offer an insight into how 'hands and feet free' driving may impact upon the interactions that take place across the driving system. Figure 3.6 represents Level 10 automation (full automation) of the pedestrian detection task. Unlike Level 8 automation, Level 10 automation sees the monitoring roles between the driver and the automated system become reversed. Rather than the Pedestrian AEB being responsible for monitoring the behaviour of the driver, the driver's primary role within the driving system at Level 10 is to monitor the behaviour of Pedestrian AEB and the other systems in which the automation interacts. A fully automated pedestrian detection subsystem would essentially delegate the task of 'braking' and 'steering' in emergency situations to automated subsystems. The driver would essentially become a monitor of vehicle subsystem behaviour. However, in contrast to Endsley and Kaber's (1999) taxonomy, in which full automation signals that the human operator is not able to intervene, this concept is currently less pertinent to driving because the driver remains the key actor in overall system safety (Brookhuis et al., 2003; SMART, 2010). Driver intervention at this level of automation may be caused due to a conflict in subjective imminent hazard perception. Where Pedestrian AEB may deem it safe for the vehicle to continue on its trajectory, the driver may wish to intervene. In this way, it seems that Pedestrian AEB could increase driver workload in unintentional

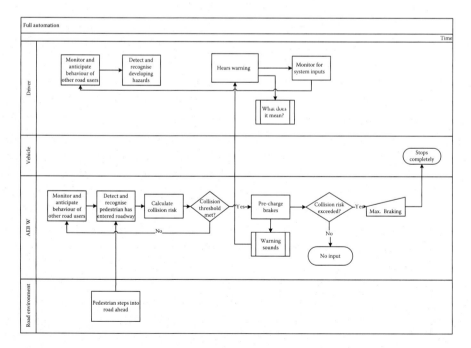

FIGURE 3.6 Hypothetical representation of full automation of pedestrian detection (Level 10).

ways, through increased vigilance demands, lack of feedback and a wider range of decision options (Brookhuis and de Waard, 2010; Parasuraman et al., 2000). This is because in the event that Pedestrian AEB fails to (a) detect a vulnerable road user; (b) misjudge pedestrian trajectory; (c) misjudge vehicle trajectory; and (d) travel above speed thresholds; the driver must respond accordingly to changes in their environment. If failure does occur, driver vigilance of system state will play a key role in whether or not automation failure will be overridden successfully. The driver must maintain an awareness of how the system is behaving and continue to monitor their surroundings to ensure that the Pedestrian AEB is effectively detecting and recognising potential hazards. However, with Pedestrian AEB completing all information-processing tasks independently, the driver may become vulnerable to boredom and/or fatigue, resulting in episodes of mental underload (Dehais et al., 2012; Young and Stanton, 2002). One approach to addressing the disintegration of control-feedback loops is to add a visual display to maximise the level of feedback provided to the driver. This would be in an attempt to keep the driver in-the-loop to some extent and promote active engagement with the subsystem.

DISCUSSION

Phase 1 of the Systems Design Framework has successfully shown how the role of the driver within an automated driving system can be acknowledged and further explored using the principles of Distributed Cognition. Although modelling exercises like this provide a useful insight into how introducing automation into the driving system may increase the complexity of the interaction that occurs within it, further investigation is required to establish 'how' this complexity is managed. In order to validate and extend the representation of these OSDs, it is essential that the assumptions made in them are experimentally tested through the use of user trials (i.e. Phase 2 of the Systems Design Framework). Even so, this chapter has shown that OSDs are a valuable development tool that can offer a quick and effective way of visualising how a technology may impact upon overall system operation (Wallace et al., 2000). Although no one method is capable of representing the true complexity of the driving task, OSDs provide a good foundation for future investigation at the very early stages of system development.

Of course, with a rapidly changing and unpredictable environment, the interactions that take place are likely to be highly adaptable and not constrained to the processes outlined in this chapter and representational methods. Constraining complex behaviour into 'swim lanes' is of course limiting, yet offers reasonable approximation of Distributed Cognition within the driving system much like how Sorensen et al. (2011) dissected a cockpit environment. Although specific functionality issues were not represented due to its exploratory use of OSDs in representing Distributed Cognition in driving, as long as functionality issues are considered, OSDs may prove to be a useful human–machine design and allocation of system function tool in the development of future automated systems. What these methodologies demonstrate is that ironically, driver task loading does not appear to reduce as the level of automation increases. Quite possibly workload will actually increase as the driver is required to monitor and anticipate the road environment, the behaviour of other road users in addition to the automated aspects of vehicle control synthesising the wider literature on malleable attention (Young and Stanton, 2002).

FUTURE DIRECTIONS

System models are an invaluable resource but user trials are needed to validate them. With one of the greatest challenges remaining to be addressed surrounding whether or not the introduction of higher level automation into the driving task brings about any performance increments or decrements on behalf of the driver, Chapter 4 focusses upon investigating the interaction that takes place between the driver and other system agents. This requirement is essential to ensure that automation implementation optimises overall system performance representing the application of Phase 2 of the Systems Design Framework.

4 Exploring the Use of Verbal Protocol Analysis as a Tool to Analyse Driver Behaviour

INTRODUCTION

In order to further improve and extend our understanding of Distributed Cognition in driving, the Systems Design Framework encourages researchers to make use of extension methodologies to further develop OSD representations (see Chapter 3 for examples). While OSDs can model the behaviour and interaction that may occur between system agents, VPA (Ericsson and Simon, 1993) can offer a unique insight into the cognitive aspects of driver behaviour that could serve to validate the assumptions made within these visual representations. The purpose of this chapter is to identify processes and procedures that may be useful for future research in developing our understanding of the driver's role within the driving system. Chapter 4 therefore serves to provide a link between Phase 1 (Modelling) and Phase 2 (Validation) of the Systems Design Framework as highlighted in Figure 4.1.

ANALYSING VERBAL PROTOCOLS FROM DRIVERS

Verbal reports offer a way to record the human thought process and are a key element in the analysis of decision-making (Ericsson and Simon, 1993). They have been widely used in a number of domains including nursing (e.g. Hoffman et al., 2009; Whyte et al., 2010), driving (e.g. Lansdown, 2002; Walker et al., 2011), software engineering (e.g. Hughes and Parkes, 2003) and the nuclear industry (e.g. Lee et al., 2012). When carried out using Ericsson and Simon's (1993) technique description, Russo et al. (1989) suggested that a rich dataset can be generated. In an extensive review of VPA methodology, Ericsson and Simon (1993) suggested that verbalisations can give insight into the contents of a person's working memory. Verbal reports therefore offer a way to reveal what a driver is thinking when they are completing a task which provides real-time insight into the information that is used and the mental processes that are applied during the decision-making process (Hughes and Parkes, 2003).

There are two primary methods of VPA: concurrent think aloud and retrospective think aloud (Ericsson and Simon, 1993). Concurrent methods require the participant to provide verbal reports during a task while retrospective reporting is conducted immediately after a task is completed. While concurrent data can be very interesting,

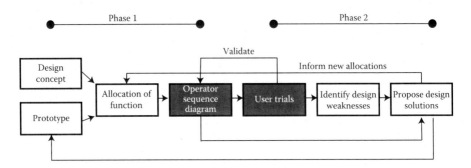

FIGURE 4.1 Aspects of the Systems Design Framework focussed on this chapter (as highlighted in grey).

Bainbridge (1999) argues that it still remains incomplete. Just because the subject fails to report something does not mean that they are unaware of it, or know about it. For this reason, retrospective verbal reports can be useful in supplementing concurrent protocol data by enabling direct exploration of the subjects' knowledge (Bainbridge, 1999).

Concurrent verbal protocols are an attractive methodology as it is unobtrusive, and from an information-processing perspective, concurrent verbalisation requires 'inner language' to be expressed from short-term memory (Taylor and Dionne, 2000). This direct style of reporting is thought to elicit valid and reliable data as individuals are not able to edit and change their responses (Ericsson and Simon, 1993). However, concurrent verbal protocols are increasingly prone to reactivity resulting in either enhanced performance (due to a more structured approach to work) or poorer performance (resulting from changes to workload patterns) (Russo et al., 1989).

In contrast, retrospective recall can decrease reactivity (van den Haak et al., 2003). This is because drivers are given the opportunity to complete a task in their own time and performance is likely to be consistent with normal driving. Even so, this style of reporting should be used only for short task durations as lengthy tasks carry an increased risk of omission and constructions (van Gog et al., 2009). This could lead to increased rationalisation or false reporting (Taylor and Dionne, 2000). According to Ericsson and Simon (1993), there are a number of measures that can be taken in an effort to encourage valid and complete accounts. These include collecting retrospective accounts immediately following task completion and emphasising the importance of accuracy in reporting.

Many studies have combined the use of concurrent and retrospective verbal protocols and found them to enhance the reliability and validity of the data collected (see Taylor and Dionne, 2000 for a review). A combined approach provides insight into both the strategies used in problem-solving and also the types of knowledge used to underpin response execution. This makes a combined approach in automated driving research very appealing, especially given its potential to generate new insights into driver–vehicle interaction at increasing levels of autonomy.

Regardless of approach, there are three levels of verbalisation that Ericsson and Simon (1993) describe: first, simple vocalisation sees a reproduction of information directly (e.g. Pennington et al., 1995); second, additional translation of the thought process through recoding reflects information that is reported from a different modality (e.g. information from other modalities such as vision); and third, an explanation of thought which requires further processing effort.

Think-aloud verbalisations at Levels 1 and 2 should not change the structure of thought processing if detailed instructions are used (Ericsson and Simon, 1993). The study by Banks et al. (2014b) presented in this chapter made use of verbalisations at Levels 2 and 3. Concurrent verbalisations were most likely to produce verbalisations equivalent to Level 2 whereas retrospective verbalisations would probe for explanation of thought and therefore likely to produce Level 3 verbalisations (e.g. van der Veer, 1993). It was anticipated that this mixed methods approach would provide a useful insight into what the driver was doing while navigating through a journey in addition to the insights that were provided by quantitative performance data.

SYSTEMS DESIGN FRAMEWORK PHASE 2 – AN EVALUATION

Banks et al. (2014b) used direct observation methods to capture the underlying processes that mediate driver behaviour in emergency situations. Approaching systems design in this way enables a deeper form of analysis capable of delivering a more informed insight into how drivers use information from the environment to guide their behaviour and how drivers may experience a journey differently. This study formed part of a pilot for a much larger investigation into a pedestrian detection system following on from Chapter 3. Its primary aim was to explore and evaluate the use of VPA in the analysis of driver behaviour by highlighting its potential to enhance quantitative datasets in revealing the effects of automation on driver behaviour using the Southampton University Driving Simulator.

The Southampton University Driving Simulator was a fixed-base Jaguar XJ saloon linked to the STISIM Drive M500W Wide-Field-of-View System with Active Steering. It had three driving displays allowing for a 135-degree driver field-of-view. Using in-vehicle driving controls and high-resolution digital sensors, the driving simulator automatically recorded driving performance measures including speed, lateral position and headway.

METHOD

Participants

Three participants aged 23 (Driver 1), 24 (Driver 2) and 24 (Driver 3) were recruited to take part in this study for an in-depth analysis. Participants held a full U.K. driving licence for 6, 7 and 8 years, respectively. Participant age and level of driving experience was not considered to be a critical variable of manipulation as the primary aim of the study was to evaluate the use of VPA. Treating responses as individual case studies enabled the analysis of specific momentary behaviour in a singular manner (Hancock, 2003). Miller et al. (2002) recognised that individuals, and therefore individual drivers, perform their own personal, complex computations of cognition. Combined with VPA, this individualistic approach attempted to identify how drivers were using information to guide their behaviour. In this way, disparities in information processing could be highlighted.

Ethical permission to conduct the study was granted by the Research Ethics Committee at the University of Southampton.

Experimental Design and Procedure

This study used one simulated driving condition designed to simulate the Euro NCAP testing procedures for Pedestrian Protection Systems. In line with the known functional limitations of these systems, the experimental driving scenario was restricted to an urban environment with a mix of curved and straight sections of road, opposite-flow traffic and parked cars on either side of the roadway. At various points along the route, five critical braking events took place that were based upon typical circumstances surrounding pedestrian accidents (Lenard et al., 2011, 2014). For example, pedestrians of varied heights and gender crossed from the kerb side with or without obstruction but required the driver to take intervening action in order to avoid a collision with them. The simulated travel speed of the critical pedestrian events was set to 4 feet/second as per the examples provided in the STISIM manual when the driver was within 30 m of the pedestrian on the side of the road. The travel speed of the pedestrian ensured that the driver needed to take evasive action. For every one critical event, there were three non-critical events (Lees and Lee, 2007; Parasuraman et al., 1997) that consisted of pedestrians crossing the road ahead at a much greater distance. These non-critical pedestrian events were defined as pedestrians who were set to enter the vehicle path at 4 feet/second when the driver was within 197 feet of them. This means that in each driving condition, drivers were presented with 5 critical braking events and 15 non-critical braking events. The non-critical events did not require driver intervention.

Upon providing informed consent, participants received a 30-min training session to familiarise themselves with the simulator controls and received training in providing verbal commentaries. All participants watched a short video clip offering an example of concurrent verbal protocol and listened to another example using an audio recording. This was considered to be an appropriate approach to training as learning by observation is a recommended learning technique (Bandura, 1986). Cognitive load research has indicated that being shown a worked example can prove to be very effective for novices (van Gog et al., 2009), and other research has used similar video-based training that allows participants to observe an expert completing a task while concurrently verbalising their thoughts (e.g. Wouters et al., 2008). Verbalisation during the simulated driving task involved the driver talking aloud about their thoughts, representing Level 2 of the Ericsson and Simon (1993) taxonomy, while a retrospective report was designed to elicit Level 3 verbalisation (explanation of thought processes) following task completion.

Following this induction, drivers were instructed to drive along the pre-defined route as described above and verbalise their behaviour. Verbal commentaries were recorded using a digital recorder and microphone in both concurrent and retrospective reporting. If verbal commentaries stalled, prompts were given to encourage the driver to continue talking for the duration of the simulated driving condition (approximately 10 min). At the end of the simulated driving phase, drivers were then

TABLE 4.1

CDM Probes Used by Analyst to Generate Insight into Driver Decision-Making

Probe	Associated Questions
Situation Assessment	Can you summarise what happened in these emergency events?
Cues	How did you know that a critical situation was occurring? What did you see or hear?
Generalisation	Were you reminded of any previous experiences in which a similar decision had to be made?
Basis	Can you tell me what strategies you used to avoid the pedestrian? What information did you use in making this decision and how was it obtained?
Options	Were there any alternative strategies you could have used, or thought about using? Why were these rejected?
Time Pressure	How much time pressure was involved in your decision-making?

asked to complete a retrospective verbal commentary from memory, enabling the analyst to probe for further information using modified exemplar questions from the CDM (Klein et al., 1989; Table 4.1).

Although it is acknowledged that the act of producing a concurrent verbal commentary may interfere with response execution in the driving task (e.g. it may increase reaction times and affect braking behaviours, Ericsson and Simon, 1993), the primary purpose of verbal commentary collection in this study was to extend our understanding of driver behaviour. At this stage, the research was specifically interested in 'what' information drivers were using to guide their behaviour. It was not therefore a study assessing driver ability. In addition to the collection of verbal reports, the Southampton University Driving Simulator recorded data relating to braking and steering inputs so that the analyst would be able to compare the verbal reports to what actually happened during the simulation.

Data Reduction and Analysis

Verbal reports were initially transcribed and then segmented into identifiable units of speech. An initial coding scheme based upon the information-processing functions outlined by Endsley and Kaber (1999) and later adapted by Banks et al. (2014a) was used to analyse the content of verbal commentaries. These functions included Monitor, Anticipate, Detect, Recognise, Select, Decide and Respond. Refinement of the coding scheme followed using a hybrid of top-down (theory-driven) and bottom-up (data-driven) approaches. The iteration process was repeated until the protocols were judged to be adequately categorised into the coding scheme. Four of the initial codes were utilised (Monitor, Anticipate, Detect and Response). The authors judged that 'Detect' and 'Recognise' were too similar for further analysis as verbal reports suggested that these functions occurred simultaneously. For example, 'there is a pedestrian walking out' indicated that the driver was both detecting and recognising the hazard to be a pedestrian. For this reason, 'Hazard Detected' was included in the

TABLE 4.2

Coding Scheme for Concurrent Verbal Commentaries (Level 2) Including Description and Examples

Code	Description	Examples
Monitor	Description of the route and the information used	'In the distance there is a … light control crossing or junction which is red' 'Just checking my mirrors'
Anticipate	Statements referring to being aware of what will happen and taking action to be prepared	'Lights ahead at a junction have just turned amber so anticipating having to stop' 'Just coming into a town … so I'm going to begin to slow down'
Hazard detected	References to potential hazards within the environment	'There's a pedestrian on the side of the road' 'Pedestrians on my left hand side after the lights and parked cars either side'
Response	Statements describing what the participant is doing to cope with hazards only	'I have swerved' 'I've just had to slam on the brakes'
Justification	Statements giving a reason for a certain choice or action	'(I have swerved) to narrowly avoid them' '…wasn't in any danger in terms of my speed'
Evaluation	Statement evaluating previous actions, choices or information	'…not paying attention' 'Thankfully missed the pedestrian' 'Managed to stop in time…'
Rule-governed behaviour	Statements reflecting lawful driving	'…now waiting for the lights to change'
Interaction with vehicle	Statements relating to use of vehicle instruments (not related to intervening action) but inclusive of steering, braking and gear change	'Changed to second gear and accelerating' 'First gear, changing into second'

coding scheme. 'Select' and 'Decide' were also omitted because they failed to 'fit' any of the verbal reports. There was no evidence in the verbal reports that suggested that responses were in any way planned or options were rejected. Banks et al. (2014b) felt that 'Response' captured this more clearly. The final coding schemes for concurrent verbal commentaries contained eight categories. For descriptions and examples of these categories, see Table 4.2.

Although different categories were used for retrospective verbal commentaries due to the inherent differences in the information recorded between concurrent and retrospective recall, there was some overlap between the coding schemes (see Table 4.3). For instance, definitions for 'Evaluation' and 'Rule-Governed Behaviour' remained the same whereas 'Response' in concurrent reports and 'Reactive Response' in retrospective reports were very similar. In contrast to concurrent verbalisations that generated information regarding action, retrospective reports offered a greater insight into how decisions had been reached. Specific to retrospective reports were

TABLE 4.3
Coding Scheme for Retrospective Verbal Commentaries (Level 3) Including Description and Examples

Code	Description	Examples
Awareness	Observational statements giving insight into driver awareness	'Cues from the road and surrounding area so traffic lights and such' 'Increased risk of something happening I guess'
Knowledge-based behaviour	Statements referring to previous driving experience	'I was watching those pedestrians carefully because they were in full view and I was aware of the potential for them to walk out' '…I knew that I needed to brake as that's how you slow down'
Rule-governed behaviour	Statements reflecting lawful driving	'To navigate through the town safely and obey the laws of the road'
Strategies	Statements describing what the participant did to cope with hazards	'I tried to anticipate them approaching'
Reactive response	Statements referring to the pressure of action	'It's just one of those oh my God, I need to stop….'
Evaluation	Statement evaluating previous actions, choices or information	'There was only one issue and that was not reacting [quickly] enough'
Consideration of alternative strategies	Statements referring to possible alternative strategies	'Obviously you're trying to avoid a pedestrian and I think you have to decide whether braking or swerving will work…'
Evaluation of alternative strategies	Statement evaluating the success of alternative strategies	'I think most people would swerve because your hands are already on the steering wheel whereas your feet are concentrated on the accelerator and it may take more time to find the brake…'

'Consideration of Alternatives' and 'Evaluation of Alternatives' because concurrent reports failed to provide insight into any pre-decisional behaviour regarding choice of action. In addition, 'Strategies' was also included in the retrospective coding scheme because the verbal reports suggested that rather than driving being a completely reactive task, drivers were using strategies to guide their behaviour (see Table 4.3). Again, refinement of the coding scheme was guided by the information available in the protocols. The iteration process was repeated until the protocols were judged to be adequately categorised into the coding scheme. Wherein concurrent recall (i.e. Level 2: information from modalities) participants were able to freely discuss whatever came to mind, retrospective recall (i.e. Level 3: explanation of thought) probed for information regarding specific events within the scenario. Banks et al. (2014b) felt that it would therefore be inappropriate to use the same coding schemes to analyse the verbal commentaries.

Results

Frequency of Observations

A comparison of Figure 4.2a and b indicates that the total number of observations made during concurrent reporting was far greater than those made in retrospective reporting. This is a trend often reported in the literature because retrospective reports require the driver to retrieve information from their long-term memory (Camps, 2003; Ericsson and Simon, 1993; van Gog et al., 2005). Concurrent verbal reports seem to provide a more complete representation of cognition in real-time (Ryan and Haslegrave, 2007; Whyte et al., 2010).

It is clear that the execution of the retrospective recall failed to deliver the richness of information that was expected despite using exemplar questions from the CDM. This may be due to limitations in accessing information from long-term memory following a 10-min task (e.g. drivers were unable to recall their thoughts of individual events post-trial). Ericsson and Simon (1993) suggested that this retrieval failure results from similar memory structures for multiple events being accessed rather than the cognitive processes for a single event being recalled. It is contended that the probability of retrieval failure increases if an individual completes a series of similar problems in a short space of time (Ericsson and Simon, 1993). Even so, retrospective insights were included in this chapter because they were still very interesting.

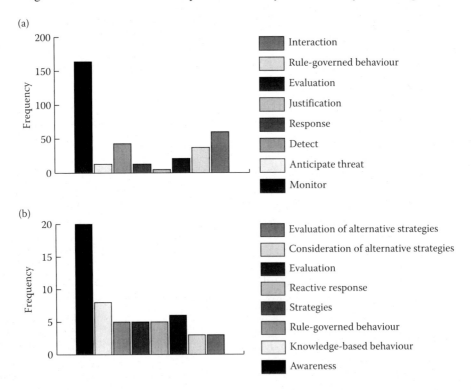

FIGURE 4.2   Frequency of code occurrence for (a) concurrent verbal reports (including both critical and non-critical events) and (b) retrospective verbal reports (critical events only).

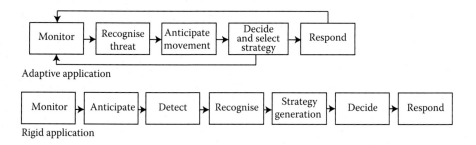

FIGURE 4.3 Adaptive versus rigid application of control elements.

Concurrent VPA suggests that driver monitoring is the overarching function that determines the pattern of subsequent processing (e.g. pedestrians who were obscured from view compromised the ability of the driver to both anticipate their movements and detect them). This supports the 'Information, Position, Speed, Gear and Acceleration' (IPSGA) system of car control approach outlined by Stanton et al. (2007a). The IPSGA system is loosely tied due to the dynamism of potential hazardous events. Each element of the IPSGA can be viewed as an underpinning behaviour of car control. Much like the researcher here, Stanton et al. (2007a) placed emphasis upon adaptive application of these system elements rather than a rigid sequence of behaviour (see Figure 4.3).

In this way, driving schemata (cf. Neisser 1967) guides monitoring behaviour that can assist in the interpretation of information presented in the wider environment and lead to subsequent action. For example,

'...just coming into a town now...so I'm going to slow down a bit...'

'Can still see a lot of pedestrians on the pavement so I'm just going to be mindful'

'...I was more cautious when I was driving through the town because there was lots going on'

'I tried to anticipate them approaching and then when I did see that someone was pulling out or crossing the road I tried to put on the brakes'

These statements show that drivers were anticipating that a hazardous situation may arise based upon the demand of the task at specific moments in time. Pedestrian detection reflects a more adaptive approach to information processing, as drivers repeatedly recognise the threat and anticipate having to respond in some way. Additionally, it would also appear that driving behaviour was also being guided by top-down influences that were external to the driving task under simulation. For example, drivers slowed for all traffic signals and appeared to have an obvious desire to maintain speed limit boundaries. Reflective statements within retrospective reports offered unique insights into the cognitions:

'I like to know how fast I'm going [because] I'm one of those people that will get caught speeding...'

Reflective statements within retrospective reports offered an important opportunity to make inferences about driver decision-making throughout the driving task. For instance, when participants were asked about strategies to avoid pedestrians during retrospection, one respondent reported (in reference to non-critical pedestrian events) the following:

'I knew I wasn't going fast enough to ever hit them…'

This statement suggests that the driver was performing some form of risk assessment that was not obvious in the concurrent report.

Extending Performance Data with Verbalisations

Despite not being able to prove or disprove what a driver is actually thinking, it is possible to relate the verbal reports collected as part of this study to a step-by-step breakdown of the built scenario and performance data generated by the Southampton University Driving Simulator. Driving performance data recorded by the simulator are presented in Figures 4.4 through 4.6. These figures represent the speed and braking profiles of the three participants with annotations taken directly from their associated verbal reports. Responses to critical braking events are characterised by peaks in longitudinal acceleration due to braking. Other braking events throughout the journey can be attributed to red traffic signals that were included for realism. Logged pedestrian collisions indicate that all three drivers were involved in at least

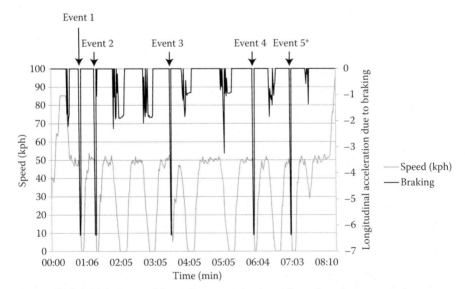

Event 1: 'Pedestrian has entered the carriageway and we've performed an emergency stop'
Event 2: '…we've just narrowly missed the pedestrian'
Event 3: 'I have swerved to narrowly avoid them and also broken'
Event 4: '…no perceived threats'
Event 5: 'Pedestrian was seen but the vehicle has not stopped in time'

FIGURE 4.4   Driver 1 profile of speed and braking behaviour (* pedestrian collision).

one collision throughout the experiment. Using hard data alone is however unable to provide any more insight into 'how' or 'why' collision occurred. Therefore, the benefit of collecting verbal commentaries is that an insight into contributory factors can be attained.

The VPA relating to Figure 4.4 generated insights into possible contributory factors in the pedestrian collision both concurrently and in retrospection. Despite making five distinct braking attempts distinguished by the sharp peaks in longitudinal acceleration due to braking, one of these attempts resulted in collision with a pedestrian. Looking at the quantitative data alone does not give any insight into possible cause. After all, the peak at critical braking event 5 does not appear to differ from the others. However, the data available in the retrospective VPA suggests that driver reactions could have been slower for this particular event:

'...I did spot the hazard. I just wasn't reacting quickly enough. There wasn't a lot of time between identifying the hazard and being able to stop in time'

According to the verbal commentary relating to Figure 4.5, it seems likely that resource conflict for visual attention between the instrument cluster (Eyes-Down Display) and road scene (Eyes-Up Display) was the causal factor in the collision that occurred at critical braking event 5. This conflict may have resulted in the driver not responding to the pedestrian event as evidenced in Figure 4.6 that shows no braking

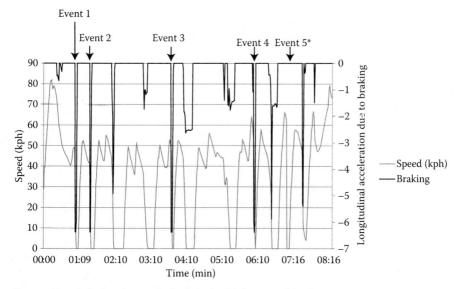

Event 1: 'I've just had to slam on the brakes. Thankfully I missed him'
Event 2: 'Just slammed on the brakes again'
Event 3: 'Pedestrian has ust walked across the road in front of me'
Event 4: 'I only just missed that pedestrian by a few, a few inches there'
Event 5: 'I'm travelling, just at 60 ... I've just hit a pedestrian'

FIGURE 4.5 Driver 2 profile of speed and braking behaviour (* pedestrian collision).

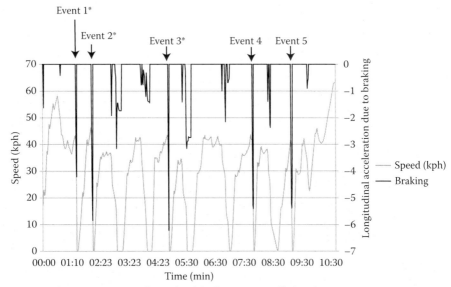

Event 1: 'I can see a crossroad sign ahead. Pedestrian just pulled out'
Event 2: 'Pedestrian pulled out...stopping. Again!'
Event 3: 'More parked cars, car just come towards me on the right hand side. Pedestrian!'
Event 4: 'Pedestrian crossing so I'm braking. Managed to save that one'
Event 5: 'Managed to stop in time – she's a slow walker'

FIGURE 4.6   Driver 3 profile of speed and braking behaviour (* pedestrian collision).

response was recorded. Although we want drivers to be aware of their speed, especially in highly populated urban environments, resource conflict occurred at a critical time. With less attention being paid to the Eyes-Up Display (road environment), there was a failure to detect the pedestrian and hence resulted in a 'no response'. In this way, normal driving tasks (i.e. checking speed) appears to have been a cause of distraction for the driver. Without recording concurrent verbal reports, this information would have been lost. However, this information may be useful to inform the design of a warning system that can alert the driver to a potentially critical situation if the automation detects that the driver is not looking at the road ahead.

In contrast, the verbal commentary relating to Figure 4.6 indicates that pedestrian collisions at critical braking events 1, 2 and 3 may have been a result of inadequately controlling the braking function. A comparison of Figures 4.4 through 4.6 indicates that longitudinal acceleration due to braking in critical braking events varied greatly. Where Figures 4.4 and 4.5 show that longitudinal acceleration due to braking in critical braking events ranged between -4.92 and -6.37 m/s^2 (average -6.20 m/s^2) and -5.93 and -6.37 m/s^2 (average -6.27 m/s^2), respectively, Figure 4.6 clearly demonstrates smaller peaks of longitudinal acceleration. Typically, the profile for Driver 3 shows an average range of -4.71 m/s^2 for longitudinal acceleration due to braking. This suggests that rather than insufficient driver monitoring or inattention, pedestrian collisions were a result of failure to manipulate vehicle controls appropriately with some evidence provided during retrospection. This failure to operate the vehicle controls effectively implicates the overall design of the experiment.

DISCUSSION

This study has shown the potential of VPA to act as a useful extension methodology in the validation and enhancement of quantitative data obtained using the Southampton University Driving Simulator. The combination of VPA with driving simulation provides the opportunity to record complex behavioural responses in addition to information relating to driver–vehicle–world interactions in a much more overt manner than simulator data alone. As Underwood et al. (2011) argue, it is becoming more appropriate to include an assessment of higher level cognition in addition to perceptual–motor measures often reported in simulator studies.

Although the use of verbal reports measures are highly debated in cognitive psychology (e.g. Baumeister et al., 2007; Boren and Ramey, 2000; Ericsson, 2002; Jack and Roepstorff, 2002; Nisbett and Wilson, 1977), the method delivers a richness of information that would otherwise be inaccessible by any other form of data collection. For example, the use of eye tracking systems can reveal only visual scan patterns and foveal fixations, and although these can offer an accurate indication of where a subject is looking, they are not able to indicate if the participant is attending to an object with specific regard or thinking about something else (Lansdown, 2002). Quantitative data alone is also not enough to provide insight into the cognitive elements of the driving tasks and the researcher concludes that VPA can highlight issues surrounding driver error. In this study, pedestrian collisions were a result of reacting too slowly (Driver 1), resource conflict (Driver 2) and insufficient control manipulation (Driver 3). Without VPA, this data would not have been available.

The benefit of collecting verbal reports pre-automation is that we can begin to understand the knowledge base of the target population (van Gog et al., 2009) and use this as a tool to guide development for more effective systems design. Designing a system that is capable of addressing these individual forms of error for an identical task is certainly a long and enduring challenge for system designers.

PRACTICAL RECOMMENDATIONS FOR FUTURE RESEARCH

While it can be concluded that the use of retrospective verbal commentaries went some way in validating the data provided by the Southampton University Driving Simulator and information presented in concurrent verbal reports, it is acknowledged that a key weakness in the experimental design of this pilot study was the execution of retrospective probing. In order to address issues of retrieval failure, it is recommended that a 'freeze probe' technique is adopted. This involves freezing the simulation immediately after a critical braking event so that drivers can be probed specifically about the event that just occurred. In contrast to the Situation Awareness Global Assessment Technique (SAGAT; Endsley, 1988) that has traditionally used a blank screen when simulation has been 'frozen', Banks et al. (2014b) propose a 'pause' within simulation allowing the driver to view the scene in front of them. It is hoped that using this approach and exemplar questions from the CDM outlined in Table 4.1, it will be possible to uncover specific events and actions that led to the behavioural outcome observed for each critical braking event. This momentary pause is likely to generate more information that can be used to analyse driver decision-making in the most crucial parts of the task as they can draw upon information from the environment. 'Freeze probe' essentially allows for higher level analysis of individual critical braking events, providing a means to explore the uniqueness of decision-making in emergency events. Thus, VPA holds some potential in uncovering thought processes underlying behavioural outcomes in emergency situations.

In addition, modification of the training that participants receive was required. It seems likely that drivers would benefit from an increased induction period to familiarise themselves with the simulators driving controls. This could be achieved by exposing participants to a longer 'practise drive' where they could navigate through a mix of curved and straight sections of road to address lateral and longitudinal control. The practise drive should also expose participants to some of the typical driving events that may occur during the experimental drive (e.g. pedestrians walking across the road and oncoming traffic). Participants should be instructed to perform an emergency brake manoeuvre three times once they reach the speed limit used in the experimental drive to familiarise themselves with the sensitivity of the braking system. This is in an effort to reduce or avoid any issue with control manipulation that Driver 3 experienced. An increased induction period would also give drivers the opportunity to practise verbalising their driving behaviour more thoroughly.

To address concerns surrounding the use of concurrent verbal commentaries, it is recommended that a comparison of driving with verbal reports and driving without verbal reports is conducted. This means that drivers should be exposed to the same driving condition twice, easily achievable through looping the simulation. In the first half, participants should be instructed to verbalise their thoughts and analysts should freeze the simulation when a critical braking event occurs to conduct retrospective probing. In the second half of the driving condition, drivers should be instructed to 'stop talking' by the analyst. It is hypothesised that no significant differences in driving performance will prevail. However, this issue needs to be addressed in future work.

Finally, this study was a precursor for a much larger study into pedestrian detection and could therefore be criticised for its lack of data and insights. However, this study was an exploratory investigation to see whether or not VPA was an appropriate technique to analyse driver behaviour. In this way, the study has proved to be invaluable and a number of recommendations have been highlighted. Future work should make use of a larger sample size with a greater age range.

FUTURE DIRECTIONS

Chapter 5 builds on the work of Chapter 4 by continuing to investigate driver decision-making in emergencies. A greater emphasis is placed upon how driver decision-making may alter depending upon the level and type of automation that is implemented into the driving system. It is hoped that the results of these investigations will reveal important Human Factors design considerations for future implementation.

5 Using Retrospective Verbal Protocols to Explore Driver Behaviour in Emergencies

INTRODUCTION

Although automated assistance in driving emergencies aims to improve the safety of our roads by avoiding or mitigating the effects of road traffic accidents, the behavioural implications of such systems remain unknown. While steps have been taken earlier in this book to model system behaviour following the introduction of a Pedestrian AEB system via the application of Distributed Cognition to driving emergencies, the purpose of this chapter is to implement Phase 2 of the Systems Design Framework to highlight issues within design strategies used for other AEB systems, as highlighted in Figure 5.1.

This chapter explores how the level and type of automation affects driver decision-making and subsequent responses to critical braking events using evidence acquired from driver verbal commentaries based upon Banks and Stanton (2015a). Commentaries were subjected to an extensive thematic analysis and subsequent network analysis, which represents a novel approach to analysing qualitative data sources.

MODELLING DECISION PATHWAYS USING VPA

It has long been suggested that driver decision-making is a hierarchical paradigm that takes into consideration strategic decisions (e.g. route planning), tactical decisions (e.g. manoeuvring the vehicle) and operational decisions (e.g. executive acts) (Hollnagel et al., 2004; Michon, 1985). However, these high-level descriptions fail to explicitly describe the processes underlying decision-making. In order to understand how the introduction of automation into the driving task may affect the driver decision-making process, more research is needed that looks specifically at the underlying processes that mediate behaviour. For example, Endsley and Kaber (1999) assigned the following information-processing functions

- System monitoring
- Strategy generation
- Decision-making
- Response execution

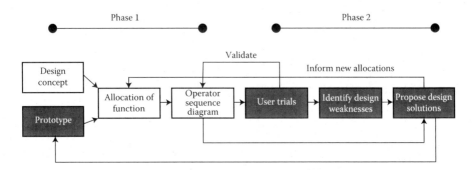

FIGURE 5.1 Aspects of the Systems Design Framework focussed on this chapter (as highlighted in grey).

to either the human operator with automated aid or as single entities. These were further adapted by Banks et al. (2014a,b) to make them more applicable to driving by incorporating an anticipatory and recognition phase within information processing. The anticipatory phase takes into consideration the feed-forward mechanisms that occur within the driving task, which are likely to be based upon previous experience while recognition provides a distinction between object detection and identification.

The study presented in this chapter uses network analysis to interrogate retrospective verbalisations (Ericsson and Simon, 1993) to investigate how automation implementation using different design strategies may affect driver decision-making and subsequent responses to critical braking events. Network analysis and its associated analysis metrics are a potentially powerful technique to use in Systems Ergonomics due to their potential to explore network resilience in the design of anticipated networks in new systems (Stanton, 2014a,b). For example, if the links between information-processing functions in driver decision-making are weakened or become severed as a result of automation implementation, it could signal that information processing may be less efficient. While network analysis metrics have traditionally been used in the analysis of social networks (e.g. Driskell and Mullen, 2005), Houghton et al. (2006) suggest that the tool can also be used to investigate decision-making and the spread of information within a system.

The main purpose of this study was to compare and contrast the processes underlying driver decision-making in emergencies at different levels of automation. A basic model of driver decision-making is presented in Figure 5.2 and uses the concepts derived from Endsley and Kaber (1999), Parasuraman et al. (2000) and Banks et al. (2014a,b) into the allocation of system function between human operators and automation with the addition of task-relevant concepts relating specifically to automation (e.g. reliance on automation and recognition of automation engagement). Data generated from a large-scale driving simulator study using the Southampton University Driving Simulator were used to validate and test the assumptions of this basic model. The solid lines within the model presented in Figure 5.2 represent expected links between information-processing functions whereas the dashed lines represent expected new links within driver decision-making as a result of

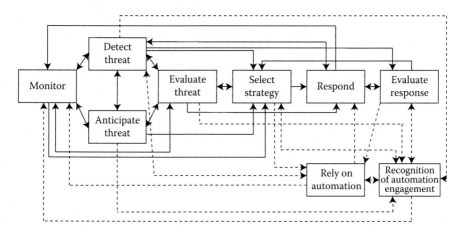

FIGURE 5.2 Proposed model of driver decision-making. Solid lines represent expected links while dashed lines represent new links afforded by automation implementation.

introducing automated assistance into the driving task. Of course, the authors acknowledged that there may also be a removal of links between information-processing nodes as automation is introduced into the driving task. If this occurred, this would be represented by the removal of links resulting in a severed network. It was proposed that evidence within driver verbalisations would enable the analyst to validate and quantify the links between information-processing functions within driver decision-making.

METHOD

Participants

A total of 48 participants were recruited from the University of Southampton student and staff cohort. All participants held a full U.K. driving licence for a minimum of 1 year and were between the ages of 18 and 65. This was to ensure that the performance decrements associated with older drivers (i.e. over 65s) and novice drivers (i.e. drivers with less than 12 months driving experience) did not affect the results of the study.

Ethical permission to conduct the study was granted by the Research Ethics Committee at the University of Southampton.

Experimental Design and Procedure

This study along with the practical recommendations outlined in Chapter 4 made use of the same basic experimental design. Upon providing informed consent, participants received training in the provision of verbal commentary. Following this introduction, participants were then invited to take part in a 'practise drive' in the simulator where they were asked to navigate through a mix of curved and straight road sections to familiarise themselves with the lateral and longitudinal controls. During this time, participants were provided with examples of typical driving events that they could occur during the experiment (e.g. pedestrians walking across the road and oncoming traffic). They were also encouraged to manipulate the braking system to get used to the sensitivity of the braking system in an effort to limit any interference between control manipulation and the verbalisation of their behaviour.

In total, drivers were required to complete four experimental drives that were designed to reflect three different levels of automation. These took into consideration the alternative methods of systems design implementation in the pedestrian detection task and were chosen for both their symmetry within Endsley and Kaber's (1999) taxonomy and relevance to the task:

1. *Manual (Level 1)*: A non-automated drive required the driver to complete all of the physical and cognitive tasks associated with driving without assistance.
2. *Decision Support (Level 4)*: A warning made up of a visual (head-up) and auditory signal that could alert the driver to hazardous situations within their environment was developed to reflect Level 4 of Endsley and Kaber's (1999) taxonomy.
3. *Automated Decision-Making (Level 8)*: An escalating warning approach that uses a warning to try and alert the driver to a critical hazard in the road ahead coupled with autonomous braking (AEB W).
4. *Automated Decision-Making (Level 8)*: A non-warning-based approach (AEB nW) that deliberately omits the use of a warning to alert the driver prior to autonomous braking.

Aside from the manual driving condition that required drivers to complete all of the physical and cognitive tasks associated with driving, the remaining three automated driving conditions (warning, AEB W and AEB nW) were designed so that drivers received assistance in the intervention of critical emergency events. These were defined as any event that without intervention would result in an accident. All three systems were designed using Visual Basic and were built to intervene in all critical emergency events regardless of driver intervention using a simple timing analysis (Table 5.1). This means that once active, AEB input would override any driver input. However, drivers were encouraged to respond as they normally would regardless of automated assistance being present. Throughout the experiment, drivers were instructed to try and maintain a speed of 50 kph (approximately 30 mph), which is consistent with city driving in the United Kingdom.

In order for drivers to understand the functionality of each of these systems, participants were instructed to complete mandatory practise drives before each condition. It was made clear that these systems would only intervene in critical emergency events and would be triggered autonomously. Trials were counterbalanced to remove order effects.

The basic driving scenario for all driving conditions was identical to the one outlined in Chapter 4 with practical recommendations considered. Again, there were 5 'critical braking events' and 15 non-critical braking events within each experimental drive. 'Critical' event locations were altered in an effort to reduce learning effects. Each experimental condition lasted for 10 min, which although represents

TABLE 5.1
Automation Design Analysis

Speed (kph)	Speed (m/s)	Distance to Hazard on Warning Trigger (m)	Distance to Hazard on AEB Trigger (m)	TTC Warning Trigger (s)	TTC AEB Trigger (s)
10	2.78	4.55	2.84	1.64	1.02
15	4.17	7.12	4.54	1.71	1.09
20	5.56	9.88	6.48	1.78	1.17
25	6.94	12.83	8.65	1.85	1.25
30	8.33	15.97	11.05	1.92	1.33
35	9.72	19.31	13.69	1.99	1.41
40	11.11	22.84	16.57	2.06	1.49
45	12.50	26.56	19.68	2.13	1.57
50	13.89	30.48	23.02	2.19	1.66
55	15.28	34.59	26.60	2.26	1.74
60	16.67	38.89	30.42	2.33	1.83

approximately half of the average journey time in England (e.g. National Traffic Survey; Department for Transport, 2013) reduces the risk of verbal omissions in retrospection (van Gog et al., 2009).

Throughout the experiment, participants were invited to verbalise their behaviour, and to overcome issues of retrieval failure in retrospective probing, the simulation was paused directly following each critical braking event in an effort to generate greater insight into the most crucial parts of the decision-making task. Using exemplar questions from the CDM (Klein et al., 1989), it was hoped that such a strategy would be better able to explore the uniqueness of decision-making in emergency events and how this may change as automation is introduced into the task.

Data Reduction and Analysis

Verbal transcripts were initially transcribed and then segmented into units of speech relating to critical braking events. An initial coding scheme was devised using a hybrid of top-down and bottom-up processes (Table 5.2).

Units of speech were coded in a mutually exclusive and exhaustive way. This means that more than one theme could be applied to each data unit if appropriate (where one unit equalled one sentence). Only the first seven themes were relevant to the manual control driving condition as the final two were only relevant to driving conditions where automated assistance was present. Five complete verbal transcripts (i.e. verbalisations relating to all four driving conditions) were selected at random and subject to further analysis by a secondary coder to calculate inter-rater reliability. A score of 90% agreement was achieved between coders, meaning that the codes applied by the primary coder were accurately matched by the secondary coder 90% of the time.

TABLE 5.2

Coding Scheme and Examples Used to Analyse Retrospective Verbal Reports

Code	Definition	Example
Monitor	Acquire information from the environment relevant to perceiving system status	'Well the street scene widened out and I noticed that this pedestrian was standing back from the white line.'
Detect threat	Recognition of potential hazards within the environment	'But there was also another pedestrian a bit beyond him....'
Anticipate threat	Being aware of what will or may happen in the future	'I thought he was going to cross....'
Evaluate threat	Statement evaluating the risk that perceived threats may hold	'...I kind of evaluated them as a threat and then took action based on that...'
Select strategy	Deciding on a particular option or strategy	'...I had a bit more time to brake'
Respond	Implement chosen strategy *or* react to situation, e.g. braking, steering	'I braked...'
Evaluate response	Statement evaluating previous actions or choices	'...I felt I did enough, whether there was enough force on the brake I'm not 100% sure, but I felt I saw a lot quicker'
Recognition of automation engagement	Statements reflecting an awareness of the presence of automation while driving	'...the first thing I saw was the warning, actually, rather than the pedestrian...'
Rely on automation	Conscious decision to rely on automation during emergencies	'I decided not to brake. I wanted to test out the system...'

RESULTS

Frequency of Links between Processing Nodes

The decision-making pathways within driving emergencies (i.e. the connections between information-processing functions) were defined as the sequential paths of coding within the verbal transcripts. For example, references to driver 'monitoring' in the verbal commentaries may have been succeeded by references of driver 'anticipation', and this would represent a link between the 'monitor' and 'anticipate' nodes within the decision-making model. This may have then been superceeded by references to 'detecting' a threat, and this would represent a link between the 'anticipate' and 'detect' nodes within the decision-making model. A total of 960 emergency situations were analysed in this way, with 240 in each driving condition.

While in its simplest form, a task network such as that presented in Figure 5.2, can be used to represent how nodes are connected, the network can also be represented as a matrix of association, meaning that quantitative metrics can be applied directly to the data (Houghton et al., 2006). With this in mind, the results of the analysis are presented in four matrix of association tables (reflecting each driving condition) to enable further analysis. Table 5.3 shows the frequency of sequential coding in

TABLE 5.3

Matrix of Association Showing Frequency of Links between Information-Processing Nodes in the Manual Driving Condition

	Monitor	Detect Threat	Anticipate Threat	Evaluate Threat	Select Strategy	Respond	Evaluate Response
Monitor		278	31	4	12	0	0
Detect threat	22		53	44	100	86	0
Anticipate threat	15	12		11	48	0	0
Evaluate threat	36	4	3		16	3	0
Select strategy	0	0	0	1		178	0
Respond	16	10	0	2	0		25
Evaluate response	0	0	0	0	5	2	

manual driving, while Tables 5.4 through 5.6 show the frequency of sequential coding in the warning, AEB W and AEB nW conditions, respectively.

Network Analysis

In order to understand the differences in network dynamics of individual decision-making models (i.e. four to represent the different levels of automation used within this study), network density, diameter, cohesion and sociometric status were calculated. Table 5.7 presents the results of these calculations that were generated using the Applied Graphic and Network Analyses package (AGNA™, version 2.1; Benta, 2005). All four of these networks can be described as weighted (i.e. non-uniform) and non-symmetric (i.e. directed). As to be expected, the size of the network increased as the level of automation increased as evidenced by the increased number of nodes and edges.

It indicates that the manual model is the most densely connected network whereas the warning model represents the least densely connected. It seems likely that the effectiveness of the decision-making process will be based upon strong interconnectedness between information-processing nodes, meaning that these findings suggest that a manual model is the most desirable system network to adopt. However, with mandatory fitment of AEB (European Parliament and the Council for the European Union, 2009) over the next few years, it would appear that the most desirable system configuration to adopt is AEB W as this represents the strongest interconnected network after manual driving.

Table 5.7 indicates that less 'hops' were required to get from one side of the network to the other within the manual model, meaning that network diameter was smallest for manual driving. However, the introduction of automation into the driving task led to a slightly increased number of 'hops', meaning that additional complexity has been introduced into driver decision-making and that the links between information-processing nodes are weakened due to the size of the overall network.

Evidence from driver verbalisations suggests that the manual model has more reciprocal links between information-processing nodes in comparison to all other models, showing that it is the most cohesive (Table 5.7). The dynamism of the

TABLE 5.4

Matrix of Association Showing Frequency of Links between Information-Processing Nodes in the Warning Driving Condition

	Monitor	Detect Threat	Anticipate Threat	Evaluate Threat	Select Strategy	Respond	Evaluate Response	Recognition of Automation Engagement	Rely on Automation
Monitor		214	12	0	0	0	2	44	0
Detect threat	20		59	23	58	80	0	27	0
Anticipate threat	7	7		2	52	0	0	6	0
Evaluate threat	36	0	2		10	0	0	11	0
Select strategy	0	0	0	0		160	0	1	1
Respond	0	0	0	0	0		0	1	0
Evaluate response	0	0	0	0	10	0		0	0
Recognition of automation engagement	19	49	0	0	28	16	0		0
Rely on automation	0	0	0	0	0	0	0	0	

TABLE 5.5

Matrix of Association Showing Frequency of Links between Information-Processing Nodes in the AEB W Driving Condition

	Monitor	Detect Threat	Anticipate Threat	Evaluate Threat	Select Strategy	Respond	Evaluate Response	Recognition of Automation Engagement	Rely on Automation
Monitor		157	13	1	0	0	0	64	0
Detect threat	2		34	33	67	66	1	29	0
Anticipate threat	0	9		2	35	0	0	4	0
Evaluate threat	0	1	3		12	1	0	16	0
Select strategy	0	0	0	0		131	0	0	34
Respond	0	3	0	0	0		13	3	0
Evaluate response	0	0	0	0	11	0		0	4
Recognition of automation engagement	3	71	0	0	35	0	2		2
Rely on automation	5	0	0	0	0	1	0	1	

TABLE 5.6
Matrix of Association Showing Frequency of Links between Information-Processing Nodes in the AEB nW Driving Condition

	Monitor	Detect Threat	Anticipate Threat	Evaluate Threat	Select Strategy	Respond	Evaluate Response	Recognition of Automation Engagement	Rely on Automation
Monitor		229	9	1	4	0	0	10	0
Detect threat	0		56	30	91	44	1	24	10
Anticipate threat	2	7		6	49	0	0	0	0
Evaluate threat	0	0	3		20	3	0	1	0
Select strategy	0	0	0	1		163	0	2	42
Respond	3	1	0	0	0		20	26	0
Evaluate response	0	0	0	0	17	0		2	27
Recognition of automation engagement	3	10	0	0	27	6	8		2
Rely on automation	5	0	0	0	0	0	0	1	

TABLE 5.7

Contrasting Network Metrics for Different Decision-Making Models Based upon AGNA™ Analysis

	Manual	Warning	AEB W	AEB nW
Number of nodes	7	9	9	9
Number of edges	31	34	43	41
Network density	0.74	0.47	0.60	0.57
Network diameter	2	4	3	3
Network cohesion	0.62	0.22	0.36	0.36
Number of links	1017	964	886	958

networks appear to change dramatically resulting in less reciprocal decision-making models, suggesting that the links between information-processing nodes are significantly weakened as a direct result of automation implementation. This means that automation may lead to a noticeable change in the driver's monitoring and response strategies, which can be seen clearly in the frequency data.

The sociometric status metric was used to identify key concepts within each decision-making model with results shown in Table 5.8. For each network, any value above the mean sociometric status value was identified as a key concept (Salmon et al., 2009). On the basis of this rule, three models (manual, warning, AEB nW) had identical key concepts underpinning driver decision-making. These were 'Monitor', 'Detect Threat', 'Select Strategy' and 'Respond', yet importantly the strength of these nodes as evidenced by Table 5.4 significantly differs. Again, it is the manual

TABLE 5.8

Contrasting Sociometric Status for Different Decision-Making Models

Node	Sociometric Status for Decision-Making Models			
	Manual	Warning	AEB W	AEB nW
Monitor	69.19[a]	41.00[a]	31.50[a]	33.75[a]
Detect threat	102.00[a]	67.75[a]	60.50[a]	62.50[a]
Anticipate threat	29.00	18.50	12.50	17.00
Evaluate threat	20.66	7.00	9.00	9.50
Select strategy	60.33[a]	40.00[a]	40.88[a]	52.00[a]
Respond	53.17[a]	35.89[a]	29.38[a]	33.25[a]
Evaluate response	6.33	3.88	4.38	6.38
Recognition of automation engagement	–	26.00	30.63[a]	15.38
Rely on automation	–	1.00	7.75	7.00
Mean sociometric status	48.67	26.78	25.17	26.31

[a] Denotes key system agents based upon the rule that any value above the mean sociometric status value reflects dominance. (Adapted from Salmon, P M et al. 2009. Distributed situation awareness: Advances in theory, measurement and application to teamwork. Aldershot: Ashgate.)

model that consistently results in the highest scores on this metric. The same key concepts also apply to the AEB W model in addition to Recognition of Automation Engagement, which suggests that the way in which information was being processed by the driver had changed. Even so, the node that consistently achieved highest sociometric status was 'Detect Threat', suggesting that this is the key node within all decision-making models. Its prominence within all four models is unsurprising when considering that in order to respond to an emergency situation, the driver must detect that a threat is present. However, again, Table 5.4 indicates that its prominence reduces as the level of automation introduced into the driving task increases. This may be attributed to the increased number of nodes within the system network or as a result of driver behavioural adaptation. For example, drivers may have been less likely to detect a hazard due to an increased reliance on automation functioning.

Looking at the statistics for Recognition of Automation and Rely on Automation specifically, it is clear that systems design had a direct effect on the strength of sociometric status. For example, warning and AEB W models had higher scores on the Recognition of Automation Engagement mode, which suggests that the warning aspect of the system caused more interference than the AEB nW system. Drivers may have been consciously waiting for the warning to activate, in this way using it as a tool to assist in the detection of a critical driving event. This would suggest issues of trust and complacency in system operation (de Waard et al., 1999; Larsson, 2012; Lee and See, 2004; Parasuraman and Riley, 1997). For example, if drivers perceived the warning system to be highly reliable, and indeed it was programmed to activate for all five critical events during the trial and thus deemed 100% effective, Young and Stanton (2002) suggested that drivers may not be able to monitor the system as closely as perhaps is warranted. In turn, this may delay driver response if they are waiting for confirmation of a collision risk. However, driver decision-making may also have been influenced by the very presence of automation. For example, drivers could have simply perceived a reduced need to respond during these critical events if they could sense that the AEB W system, in particular, was responding for them (Stanton and Marsden, 1996). This does not necessarily mean that drivers will relinquish all control over the braking function; they may instead brake with less effort than would otherwise be needed to cope with the situation. However, due to the technical sophistication of the STISIM M500W software and the algorithms used to create the automated systems used in this study, it was not possible to investigate this further. This is because the presence of automation within the system configuration led to a static performance level once an emergency manoeuvre was initiated (i.e. automatic braking overrode driver inputs).

DISCUSSION

These results are indicative of an 'adaptive application' of information processing (Stanton et al., 2007b) showing that decision-making is not a rigid, one-way process. The frequency data suggest that the strength of linkages between information-processing functions within driver decision-making also becomes significantly weakened as a result of automation implementation. The addition of automation also brings with it the introduction of new links within the decision-making network.

In summary, network analysis has shown to be a useful perspective in highlighting the inherent differences in driver decision-making based upon the level of automation within the driving system. From a Human Factors perspective, it is essential to understand how decisions are reached when response execution has the potential to lead to undesirable outcomes (Jenkins et al., 2011). The analysis of driver decision-making using modelling techniques and evidence from driver verbalisations suggests that while automation may not alter the decision-making pathway between initial driver monitoring and response as indicated by identical key concepts, it does appear to significantly weaken or sever the links between information-processing nodes. These weaknesses become more pronounced at higher levels of automation as both the physical and cognitive tasks associated with driving become shared between the driver and other system agents (Banks et al., 2014a). Thus, not only do weaknesses appear in the cognitive functioning of the network (i.e. processing underlying response execution), but also appear to weaken the link between the driver and vehicle systems (i.e. 'who' does 'what'; Banks et al., 2014a). For example, evidence from driver verbalisations suggested that 'Recognition of Automation Engagement' was used as a trigger for drivers to seek out road hazards, meaning that they were using automated assistance as a tool to guide their monitoring behaviour. If drivers did indeed choose to 'wait' for the automation to engage, we may not be improving safety in the way that we expect, despite the potential for significant reductions in accident statistics. This improper use of automation would be an unintended, emergent property of automation implementation within the system network, yet highlights the need for a greater appreciation and acknowledgement of the changing role of the driver to ensure that the negative effects of automation are controlled for. It seems that for the collaboration between the driver and the automation to be effective, drivers must have appropriately calibrated trust of an automated system (Lee and See, 2004; Madhaven and Wiegmann, 2007) and appropriate levels of awareness regarding its functionality.

FUTURE DIRECTIONS

Chapter 6 builds on the work of Chapter 5 by exploring the effects of systems design on measurable outcomes including frequency of accident involvement, reaction times and braking distances at increasing levels of automation within driving emergencies. With Chapter 5 revealing that driver decision-making alters depending on the level at which automation is set, it is important to understand the implications of this changing behaviour on measurable outcomes. Furthermore, in order to determine whether or not drivers were 'waiting' for AEB to activate or braking with less effort, analysis of the driving simulator data is needed.

6 The Effect of Systems Design on Driver Behaviour

The Case of AEB

INTRODUCTION

Although Retting et al. (2003) developed a number of engineering measures that sought to reduce the number of vehicle–pedestrian collisions, automated assistance in emergency situations can further improve road safety through the provision of visual and auditory warnings (e.g. Forward Collision Warnings) or through emergency brake assistance (e.g. AEB) in an effort to avoid or mitigate the effects of road traffic accidents. Autonomous braking intervention can be achieved in one of two ways: following a system warning that a collision is imminent (AEB W) or with no overt warning to the driver (AEB nW). Strategies for implementation have been interpreted by manufacturers of vehicles in different ways and depend heavily upon a number of important factors such as the number and type of sensors available on the vehicle, the decision to warn the driver, the automated logic itself that determines when braking will be initiated and so on (Lenard et al., 2014). The driving context does appear to influence the strategy for implementation, however, with city driving typically associated with an AEB nW system due to the certainty that autonomous assistance will be needed (i.e. greater proximity to hazards – Road Safety Analysis, 2013). This is in contrast to inter-urban driving that is often associated with increased fitment of AEB W systems. Although it could be argued, based upon these current design trends, that a Pedestrian AEB system is most likely to adopt a non-warning-based approach due to the propensity for pedestrians to be located in cities, it is not yet clear how such systems will be implemented. Even so, it is generally agreed that the implementation of pedestrian detection and autonomous braking systems, regardless of design, will be advantageous in avoiding pedestrian collisions commonly caused by human error (Habibovic et al., 2013) and that automated assistance in such scenarios will have desirable benefits (Searson et al., 2014). It seems likely that it is for these reasons that Pedestrian AEB has been recognised by the Euro NCAP (2013) as a critical safety system to be widely assessed and deployed from 2016 onwards (van Ratingen, 2012) with common accident scenarios providing a baseline for the construction of test protocols (e.g. Lenard et al., 2011).

EMPIRICAL TESTING OF AEB

Although for some the benefits of automation may outweigh any costs (Khan et al., 2012; Stanton and Marsden, 1996; Young et al., 2011), a considerable amount of research into vehicle automation over the past 30 years has shown that drivers do not always respond in the way that engineers anticipate to automated assistance (e.g. de Winter et al., 2014; Hoedemaeker and Brookhuis, 1998; Rudin-Brown and Parker, 2004; Stanton and Young, 2005; Young and Stanton, 2007a,b). Although in some instances driver behavioural change can be positive (e.g. if the driver is not looking at the road ahead or is distracted by other driving-related tasks such as checking speed, an auditory warning could alert the driver to a problem and trigger their response), it can also be negative (e.g. drivers may become reliant on automation functionality and fail to respond as expected). For example, Stanton et al. (2011) found that increasing the level of automation can lead to complacency while Parasuraman (2000) suggested that it can cause decreases to driver SA, which are closely related to issues of mental underload and overload (Young and Stanton, 2002). The authors argue that if we are to overcome these issues, more research is needed to ensure that undesirable behavioural adaptation does not occur (Stanton and Marsden, 1996; Stanton and Pinto, 2000), which can be achieved only if we acknowledge the new role of the driver in an automated driving system rather than purely focussing upon the efficacy of automation on frequency of accident involvement as a marker to determine if automated systems really do improve road safety.

Although Merat et al. (2014) suggested that both the level and the type of automation implemented into the driving system can have a direct effect on the driver's level of engagement, AEB systems are unlikely to lead to an increase in driver's willingness to engage in secondary tasks as much of the control over vehicle handling remains in the hands of the driver. However, sudden increases to driver workload (e.g. Jamson et al., 2013) can be detrimental to driving safety (Rudin-Brown and Parker, 2004), and it is in these situations that automation of varying levels could be of greatest assistance. Even so, despite the allocation of system function being vital in understanding how automation may change the role of the driver (Banks et al., 2014a,b), the automobile industry continues to be plagued by criticism for failing to acknowledge the changing role of the driver within automated systems once such systems have been deployed (Banks et al., 2013). This means that we do not fully understand the complexities of driver–automation cooperation (Weyer et al., 2015) and any associated performance outcomes.

Of course, the actual triggering of AEB is likely to be a rare event. However, Young and Carsten (2013) commented that AEB systems could be ripe for behavioural adaptation to occur, as drivers learn to push the limits of the system by driving more recklessly knowing that a safety net is present. However, Rudin-Brown (2010) and Hedlund (2000) propose that behavioural adaptation is less likely to occur if AEB offers no additional warning to the driver. From this perspective, a 'silent' and 'invisible' AEB system may be less prone to negative behavioural adaptation (Young and Carsten, 2013). However, it remains to be established whether or not this is the case.

METHOD

This chapter presents a continuation of the results reported in Chapter 5. However, to recap briefly, drivers were informed that the automated systems would intervene only in critical events meaning that no warnings or brake assistance would be provided for non-critical events. The AEB systems themselves were designed to intervene regardless of driver control inputs, in this way, acting as assistance if the driver initiated emergency braking first, or capable of acting autonomously if the driver failed to respond. Automatic intervention by AEB aimed to improve on the reaction time of the driver in an effort to mitigate injury rather than complete collision avoidance. Thus, drivers were told that AEB was not deemed a replacement for them. Instead, AEB should be viewed as a 'last resort' intervention strategy, in this way, maintaining the driver within the control-feedback loops for as long as possible. Any collision that occurred while AEB was active was indicative of reckless driving on behalf of the driver (e.g. elevated vehicle speeds).

RESULTS

Accident Involvement

One critical outcome measure relevant to the evaluation of automation in driving emergencies is the frequency of accident involvement. Unsurprisingly, Figure 6.1 reveals that the simulated AEB systems used in this study can reduce the overall number of accidents, regardless of design strategy. This observation suggests that a 'silent' AEB system (Young and Carsten, 2013) has no additional benefit in comparison to a warning-based AEB system within this study.

Driver–Vehicle Interaction

However, with Banks et al. (2013, 2014a,b) and others (e.g. Rudin-Brown, 2010) proposing that the design of automation may affect the way in which drivers approach

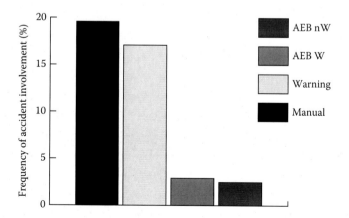

FIGURE 6.1 Percentage of drivers being involved in at least one collision across different levels of automation.

and deal with hazardous events, further analysis of the Southampton University Driving Simulator data was completed to highlight any issues associated with automated design as evidence from driver verbalisations within Chapter 5 made some suggestion that both the level of automation and the type of AEB design directly affected the way in which drivers interacted with the vehicle.

Simulator Data

At this point, the main focus of the analysis was based upon 'who' was braking during critical braking events. Only data for AEB W and AEB nW were selected at this stage as it was in these conditions that the driver could relinquish their full control of braking to the automated AEB system. Within the data files, it was clear to see 'who' (driver or automation) was responsible for initiating the braking effort as the AEB feature had a clear identifier within the dataset. This raised the possibility to essentially log 'who' interacted with the braking system first.

A chi-square test revealed that the design of the system (i.e. AEB W or AEB nW) significantly affected the frequency of 'who' (driver or automation) initiated the braking response ($\chi^2(3) = 390.29$, $p < .001$, Cramer's $V = 0.638$). Figure 6.2 indicates that AEB nW was associated with a higher prevalence of 'Driver First' responses while AEB W was associated with an increased prevalence of 'AEB First' responses. 'Driver First' in this instance simply reflects that the driver responded to the critical braking event prior to AEB activation with 'AEB First' being the opposite. Out of a total 240 critical braking events (5 per participant, per AEB condition), 34.2% ($n = 82$) 'Driver First' responses were logged for AEB W in contrast to 57.1% ($n = 137$) for AEB nW (Figure 6.2). This simply shows that different AEB design strategies do have some influence over driver–vehicle interaction patterns. This would suggest that if AEB is to remain active in the background of vehicle operation as intended, a 'silent' and 'invisible' AEB system (AEB nW) is most likely to preserve the traditional role of the driver (i.e. one that reacts prior to AEB activation). Perhaps, AEB nW in this study encouraged drivers to monitor the road ahead much more diligently than AEB W because they knew no collision warning would assist them in detecting hazardous

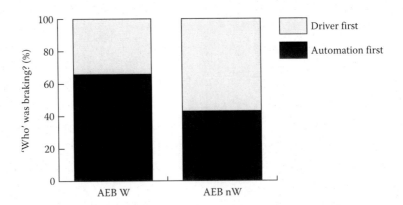

FIGURE 6.2   Graph to show 'who' (driver or automation) was braking in response to critical braking events.

situations. Either way, driver responses must have been delayed in the AEB W condition given the higher frequency of 'Automation First' responses and the fact that both AEB systems were designed using the same algorithm and thus reacted at the time to hazardous situations. With drivers being aware of the reliability of the system during simulation, it is possible that drivers were relying upon system activation, in particular, using the warning mechanism to trigger a response (similar to the purpose of a Forward Collision Warning) and thus not monitoring the environment as closely as was warranted (Young and Stanton, 2002).

Evidence from Driver Verbalisations

Further explanation for the trend presented in Figure 6.2 can be sought from additional analysis of the retrospective verbalisations collected during the study presented in Chapter 5. With the design of AEB seeming to influence 'who' initiated the braking response in critical events, the author was curious as to 'why' this happened. If, for example, drivers actively chose to delegate their control of braking to the automated subsystems (i.e. there is an unintended shift in task loading, which inhibits traditional behavioural response), a degree of skill degradation must occur (Jameson, 2003; Miller and Parasuraman, 2007; Parasuraman, 2000; Parasuraman et al., 2000; Stanton et al., 2001). Drivers may feel that the onus of responsibility for reacting to hazardous situations is now shared with an automated counterpart (Hoc et al., 2009) and thus, may delay their normal response as they attempt to cooperate with it (Hollnagel et al., 2004). Indeed, several studies have reported that performance under increased levels of automation can begin to decline as a result of lack of manual control inputs (e.g. Lavie and Meyer, 2010; Miller and Parasuraman, 2007; Parasuraman et al., 2000; Rudin-Brown and Parker, 2004; Stanton et al., 2001; Vollrath et al., 2011; Young and Stanton, 2007a,b).

Additionally, if automation is perceived to be highly reliable (as it was programmed to be in this study), drivers may not be able to monitor the system as closely as is warranted (Young and Stanton, 2002). This could delay driver response if they wait for notification or confirmation of a collision risk such as that offered in AEB W.

A matrix coding query completed using NVIVO 10 for Windows software indicated that while direct references relating to a delay in driver response to critical braking events were small (AEB W = 22 references; AEB nW = 25 references), 'waiting for automation to engage' (see code description in Table 6.1) was a conscious decision made by some participants. The following extracts are taken directly from transcribed driver verbalisations to demonstrate this idea:

'I tested the technology. Yeah, because I wasn't going quite as fast so I thought, see if it works'

'I think there's a tendency to go faster if you know something is going to catch you. So, I tend to go faster so it's like if you give someone enough rope, they'll hang themselves with it'

In contrast to Chapter 5, which reports the thematic analysis of driver verbalisations to give specific insight into driver decision-making processes, the purpose of

TABLE 6.1

Additional Codes Used to Analyse Retrospective Verbal Reports Including Frequency Counts for Different AEB Design Strategies

Code		Example	Frequency	
			AEB W	AEB nW
Thoughts on AEB	Failure to see threat	'I didn't see her at all – I was looking the oncoming traffic'	69	21
	Anticipatory behaviours	'I've spotted her and slowed down just in case'	46	63
	Alerts driver	'It helped me realise there was a problem. So I thought "I need to do something"'	61	5
	Provides comfort and reassurance	'It's like if you have a mate looking out the window for you...'	60	69
	Validates driver thought process	'...it kind of confirmed that I should have braked, braked as I could, confirmation that that was the right choice to do'	24	1
	Ignores AEB system	'I didn't even think about the automation at all'	30	55

the analysis at this stage was to support the observations of the simulator data by revealing driver perceptions of AEB. Table 6.1 defines additional codes that were used to analyse retrospective reports as well as provides examples and frequency counts.

The frequency of code occurrence for 'failure to see threat' suggests that the monitoring behaviour of drivers was affected by AEB design strategy to some extent. These failures appeared to happen more frequently when drivers were supported by AEB W and this could be partly attributed to a change in the decision-making process (as discussed in Chapter 5) as they reported less anticipatory behaviours. For example,

'I think I'm not being as cautious because I know that there is a functionality there potentially I can take maybe a bit of a wider view of the world in terms of the traffic lights or where the roads going and things like that'

Of course, drivers are unlikely to exhibit such behaviours outside the security of driving simulation. Even so, such behaviour should be considered when analysing driver–vehicle interaction patterns, remembering that humans are curious beings and will want to find and test automated system limits in some instances (Wilde, 1994; Young and Carsten, 2013).

DISCUSSION

Assuming that automation is 100% failsafe, common causes of vehicular accidents such as driver distraction, inattention and a lack of timely response could be eliminated by its implementation (Amditis et al., 2010; Cantin et al., 2009; Donmez et al., 2007; Khan et al., 2012; Stanton et al., 1997). However, this study has demonstrated that while road safety can be improved with the implementation of AEB in emergencies, the strategy of implementation determines how far the traditional role of the driver remains and this is well worth recognising.

It can be argued that in order to maximise the safety of drivers and other vulnerable road users, designers should be aware of how different design approaches at differing levels of automation could affect subsequent responses to critical hazards. This study has shown that despite asking drivers to react to hazardous events as they normally would, the implementation of AEB did affect driver–vehicle interaction not only with regards to their decision-making (Chapter 5) but also with regards to the way in which they interacted with the vehicle and braking system. This means that we may not be improving the safety of our drivers as their changing role has not been fully recognised. AEB implementation, regardless of design strategy at this stage, appears to weaken the control-feedback loops. This could lead to drivers not being equipped to cope with hazardous situations if automation failed (Sarter et al., 1997). It also implicates the concepts of trust and complacency in automation functioning (e.g. de Waard et al., 1999; Lee and See, 2004; Parasuraman et al., 1993; Young and Stanton, 2002).

It would appear that AEB implementation may lead to decoupling the link between the driver and vehicle systems within the control-feedback loops, which may explain to some extent why 'Automation First' responses occurred. This decoupling was more pronounced when AEB W was used, suggesting that a non-warning-based AEB system is better able to preserve the traditional role of the driver.

Even so, in some instances, a warning-based system may be preferable – especially in instances whereby the driver has failed to recognise a hazard in the road ahead. Table 6.1 confirms that AEB W was most capable of alerting drivers to critical situations.

Finally, it is important to remember that it is difficult to say with certainty how different levels of automation in the driving task really affect driver responses in driving emergencies due to the limitations of driving simulation. Even so, the results of this study clearly indicate that the level and the type of automation (i.e. systems design) do indeed have the potential to change the way in which the driver interacts with the vehicle (Merat et al., 2014; Stanton and Marsden, 1996; Stanton and Young, 2005).

FUTURE DIRECTIONS

The research described so far in this book has focussed on applying Distributed Cognition to existing automated architectures that are readily available to buy in the current market. The remainder of this book explores a new automation concept, one that follows the progressive pattern of automation implementation outlined in Chapter 2. Entering the developmental sphere when a product is still in the conceptual stage is a very exciting opportunity as research findings may have a significant impact on the future of its design.

Chapter 7 seeks to define the concept of 'Driver-Initiated Automation' and identify the key system agents that exist in such a system. In order to reveal the true importance of their roles within the system network, network analysis metrics will be applied to system network representations created through the application of the Systems Design Framework introduced in Chapter 3. This can be achieved by subjecting frequency counts to network analysis using Agna™ (Benta, 2005). Although Agna™ is traditionally a social network analysis tool, it can also be used for general network analysis.

7 What Is Next for Vehicle Automation? From Design Concept through to Prototype Development

INTRODUCTION

Up until now, this book has been concerned with how far existing automated technologies protect the role of the driver from negative or maladaptive behavioural change. It now moves on to looking at how future automated technologies may be deployed in the coming years. As vehicle manufacturers continue to improve the capability and sophistication of existing technologies apace, it is generally accepted by representatives within the automotive industry (e.g. Continental, 2014) that vehicle automation will continue to evolve progressively (Figure 7.1) in line with the National Highway Traffic Safety Administration (NHTSA, 2013), Bundesanstalt für Straßenwesen (BASt Expert Group; Gasser and Westoff, 2012) and the Society of Automotive Engineers (SAE, 2016) automation taxonomies. This means that both the driver and automated subsystems remain key agents within the system network and must coordinate their behaviour with one another accordingly during intermediate levels of automation. While an element of active control remains in the drivers grasp, it is critical that we understand what the driver is actually doing to ensure that they are capable of regaining control if required (e.g. recent amendment to the Vienna Convention, 1968).

This chapter investigates how multisystem automation that enables the driver to become 'hands and feet free' may affect the driving system and the role of the driver within it using Phase 1 of the Systems Design Framework (Figure 7.2). The decision to focus upon this modelling phase is based upon the discussion below.

Until the reliability of automation is sufficiently high enough to introduce fully automated vehicles onto our roads, we will remain in a state of highly automated driving, requiring the driver and automation to work cooperatively in order to maintain vehicular control (Soualmi et al., 2014). Thus, although fully automated cars are technologically feasible (Brookhuis and de Waard, 2006), during the intermediate phases of automation, the driver must remain active and in-the-loop (Hoeger et al., 2008). This poses many challenges for systems designers to ensure that the interaction between humans and automated systems are designed appropriately

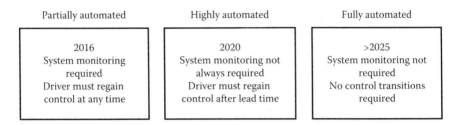

Partially automated   Highly automated   Fully automated

2016	2020	>2025
System monitoring required	System monitoring not always required	System monitoring not required
Driver must regain control at any time	Driver must regain control after lead time	No control transitions required

FIGURE 7.1 Proposed automation pathway.

(Strand et al., 2014) to ensure that the negative effects typically associated with being out-of-the-loop are minimised (Endsley and Kiris, 1995; Stanton et al., 1997; Wickens and Hollands, 2000).

With both General Motors and Nissan predicting that 'almost' driverless cars will be ready to market from 2020, it is clear that highly automated vehicles that combine multiple automated systems are coming whether we are ready for them or not. Despite the allocation of function between the driver and automated sub-systems being key in facilitating and developing driver–automation cooperation (Hoc, 2000), the industry has continued to be plagued by criticism for inadequately acknowledging the role of the driver and how it may change once these systems have been deployed (Banks et al., 2013; Stanton et al., 2007a). This means that we do not fully understand or appreciate the complexities of driver–automation cooperation in modern day cars (Stanton and Young, 2005; Weyer et al., 2015). As increased control is delegated to the automation, there is growing concern within the Ergonomics and Human Factors community that the role of the driver is not being fully recognised. A greater appreciation of the driver's ability to undertake their new supervisory role is becoming increasingly important as the average motorist becomes less actively involved in traditional vehicle handling. With out-of-the-loop performance problems and a serious concern within highly automated driving systems (Billings, 1988; Endsley and Kaber, 1997; Endsley and Kiris, 1995;

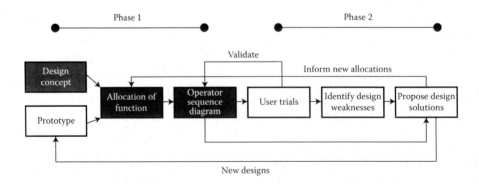

FIGURE 7.2 Aspects of the Systems Design Framework applied during this chapter as shown in grey.

Stanton and Marsden, 1996), this chapter aims to address some key research questions bearing in mind that one of the greatest challenges for systems designers is to reduce the high complexity of the automation into manageable complexity for the human driver (Kienle et al., 2009):

- What happens in an automated driving system?
- How does information flow between system agents?
- What directional flow does the information have?

The application of network analysis will show how the dynamism of the driving system and the role of the driver within it changes as more control is delegated to automated subsystems as part of the Event Analysis Systemic Teamwork framework (EAST; Stanton et al., 2013; Walker et al., 2006, 2010). The EAST framework proposes that system performance can be meaningfully described via three interlinked network representations – task, social and information (Walker et al., 2006) – to describe and analyse an activity. It was originally developed for the analysis of C4i activity (including command, control, communications, computers and intelligence) and aims to model the Distributed Cognition approach (Hutchins, 1995a) with the methodological traditions inherent in Ergonomics research (Walker et al., 2006, 2010) by enabling analysts to examine the role of actors within complex sociotechnical systems more succinctly. While task networks can be used to provide a summary of the goals and processes involved in attaining these goals, a social network can analyse the organisations of system communications or interactions that can occur between human (i.e. driver) and non-human (i.e. automation) agents (e.g. Salmon et al., 2014). EAST has been successfully applied to aviation (Stewart et al., 2008; Walker et al., 2010), rail (Walker et al., 2006), naval (Stanton, 2014a) and military C4i scenarios (Walker et al., 2006) and would appear to be an appropriate method to apply to the driving automation domain. This chapter uses the representational mediums afforded by task and social networks to describe and analyse a Driver-Initiated Command and Control System that encourages the driver to remain actively involved in the driving task much like how Baber et al. (2013) analysed the role of system agents in maritime search and rescue scenarios.

APPLICATION OF SYSTEMS DESIGN FRAMEWORK: PHASE 1

STEP 1: IDENTIFICATION OF DESIGN CONCEPT

Driver-Initiated Automation essentially enables higher levels of automated functionality but maintains the driver in-the-loop through the adoption of a command and control relationship (e.g. Houghton et al., 2006). Essentially, a system of Driver-Initiated Command and Control is analogous to a management infrastructure (Harris and White, 1987) that sees the driver and automated systems communicating and behaving cooperatively to achieve a common goal (Hoc et al., 2009). Within the driving domain, Conduct-by-Wire (Winner and Hakuli, 2006) and H-Mode (e.g. Flemisch et al., 2003) are relevant concepts relating to the design of cooperative guidance and control systems (Flemisch et al., 2014). The allocation of function between the driver and the automation should not be considered as static but instead a continual repartitioning process (Flemisch et al., 2012) where the driver and automation can influence the balance of control between system agents. For example, the driver can set higher or lower levels of automation while the automated systems can recommend or suggest tasks that can be automated (such as a lane change manoeuvre) or in emergency situations, transition control away from the driver to mitigate the effects of a collision (e.g. AEB). In this way, 'initiation' can either be prompted by the automated system itself or by the driver. The key difference in a Driver-Initiated Automation system is that the driver must 'accept' or 'ignore' requests made by the automated system. Only when the driver accepts a request can the manoeuvre be performed automatically. Thus, although continual repartitioning occurs, it is the driver who influences the balance of control. In this way, the driver is able to exercise their control and authority over the automated system, which keeps them informed of planning, directing and controlling when the resources available from the automation will be used (e.g. Builder et al., 1999). This would see the role of the driver becoming more analogous to the role of the co-pilot in aviation (Banks and Stanton, 2014; Stanton and Marsden, 1996; Young et al., 2007), meaning that although the status of the driver within the control-feedback loops of driving has changed (Banks et al., 2014a,b), the link between the driver and the vehicle is protected, to some extent, from disintegration, which is thought to lead to out-of-the-loop performance issues (Billings, 1988; Endsley and Kiris, 1995; Kaber and Endsley, 2004; Stanton et al., 1997; Vollrath et al., 2011). Although command and control sociotechnical systems are typically associated with Air Traffic Control (Walker et al., 2010) and military teams (Walker et al., 2009), there is no reason why such an approach cannot be applied to driving. After all, every 'agent' (both human and non-human) plays a critical role in the successful completion of a task (Salmon et al., 2008; Stanton et al., 2006) even when the vehicle is capable of controlling all of the physical and cognitive tasks associated with driving (Stanton et al., 1997).

When considering that each facet of technology provides a stepping stone to reach higher levels of autonomy (e.g. Banks and Stanton, 2014), it seems likely that existing automated architectures, such as cruise assist technologies (Stanton et al., 2011), could simply be extended to include lane centring and overtake capabilities in an effort to achieve higher levels of autonomy using a Driver-Initiated design approach (Figure 7.3).

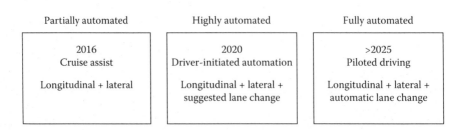

FIGURE 7.3 Hypothetical pathway for Driver-Initiated Automation implementation relating to future cruise assist technologies.

Although the combination of automated longitudinal and lateral control systems is not an entirely new concept (Stanton et al., 1997; Young and Stanton, 2002), over recent years, there has been a significant increase in manufacturers introducing their own versions that fit such a specification (e.g. General Motor's Super Cruise: Fleming, 2012). However, the future of automated highway driving, in particular, appears to point to the following subtasks of driving being completed autonomously: longitudinal control, lateral control, all round lane monitoring, lane change, merging and collision avoidance. This suggests that at some point in the future, vehicles will be capable of performing complex lane change manoeuvres independently from the driver. These technologies are likely to be developed progressively, meaning that the driver remains in ultimate control of the vehicle via a process of Driver-Initiated Automation design.

A hypothetical system such as the 2020 vision portrayed in Figure 7.3 that combines longitudinal and lateral control with a suggested lane change is likely to be classified as a system that improves the driver experience in terms of comfort rather than be marketed as a traditional safety system. Such a system will require the driver to closely monitor system behaviour in order to establish whether its performance and lane change suggestions are indeed safe to complete. This keeps the responsibility of safe vehicle operation in the hands of the driver despite the vast majority of the driving task being completely automated (Parasuraman et al., 2000). The obvious concern is that the driver will quickly become disengaged and may willingly accept that automated subsystems are operationally sound, meaning that if a lane change suggestion is offered at a time that would normally be considered unsafe by the driver (perhaps the sensing equipment fails to identify a vehicle in the adjacent lane), the driver may automatically accept the suggestion without adequately assessing the situation. Of course, the vehicle is unlikely to actually complete the manoeuvre due to the multitude of subsystem components analysing the road environment, but it could affect levels of driver trust in using such a system (e.g. Lee and See, 2004). Until such systems become available, it is difficult to say with certainty how drivers will interact with them. Even so, it is possible to model, even in the early stages of system development, how a system of Driver-Initiated Automation may be idealistically managed to ensure that the role of the driver is supported throughout active automated driving.

STEP 2: ALLOCATION OF FUNCTION

Using traditional task analysis methods (Stanton et al., 2013), it is possible to outline the processes involved in achieving a common goal. For Driver-Initiated technologies that combine longitudinal and lateral control together with a suggested lane change, there appears to be three distinct driver commands: Activate Longitudinal Control, Activate Lateral Control and finally Driver Accept/Ignore Lane Change Suggestion. The reason for these three distinct driver commands is quite straightforward; while automated longitudinal control can be used on any road type, it seems most likely that automated lateral control systems will be confined to highway driving for the time being. This is because the opportunity for lateral control to be automated is based upon the provision of clear road markings and these are not always maintained on other road types. Thus, the activation of Driver-Initiated Automation in this form is likely to follow two stages:

1. Once the driver issues the command for longitudinal control to be auto-mated, the Longitudinal Controller begins to hold, represent and modify information from the changing environment in order to reach the goal of the system network (in this case to maintain a desired speed and gap that is preset by the driver). This information is shared with the Driver-Initiated system of control and relayed back to the driver via the HMI. Of course, the driver is still free to override the Longitudinal Controller at any time by simply depressing the brake pedal. A task analysis by Stanton and Young (2005) highlighted the role of the driver within this process as one that monitors subsystem behaviour, checking that the Longitudinal Controller is maintaining the preset speed, correctly identifying vehicles ahead and responding accordingly.

2. If the driver chooses to automate Lateral Control, information is sent to the Lateral Controller relating to the driver's intention. In this instance, the shared goal of the system network is to safely stay within the confines of the lane and avoid deviation. The Lateral Controller will begin scanning the road environment for lane markings. If these are not found, the driver is notified that Lateral Control is not available via the HMI. In contrast, if lane markings are successfully identified, the vehicle can be controlled by the Lateral Controller. Again, this information is relayed back to the driver via the HMI. The system will continue to automate this task unless the system is disengaged by the driver through a steering override of approximately two Newton Metres, or the Lateral Controller fails to identify road mark-ings ahead. Again, the role of the driver would be to monitor the behaviour of the Lateral Controller, checking that the system has correctly identified lane markings and that the vehicle remains within the confines of these boundaries.

When both Longitudinal and Lateral Controllers are actively automating these driving subtasks, drivers theoretically become 'hands and feet free'. This is because they are no longer in direct control of active driving. However, they are still free to exercise their authority by overriding the system whenever they consider it neces-sary. While driving control subtasks are automated, drivers are most vulnerable to disengagement from the primary task (driving) and more likely to engage in secondary tasks (i.e. in-vehicle entertainment: Carsten et al., 2012; Jamson et al., 2013). With the likelihood of 'eyes-off-road' time increasing as the level of auto-mation increases, any failure on the part of the automation may delay appropri-ate driver response (e.g. Stanton et al., 1997; Young and Stanton, 2007b). For example, failures of an automated longitudinal control system, such as ACC, have been associated with failure to reclaim control (Stanton et al., 1997), inappropri-ate braking responses in both driving simulator studies (e.g. Young and Stanton, 2007b) and also test-track studies (e.g. Rudin-Brown and Parker, 2004). In the case of automated lateral control systems, such as Lane Keep Assist, issues relating to complacency have been highlighted (e.g. Desmond et al., 1998). These performance decrements on behalf of the driver are thought to be related to decreased SA (Endsley, 1995) and changes to driver mental workload (Stanton et al., 1997, 2001; Young and Stanton, 2002, 2007a). Reduced workload as a result of increasing levels of automation (Young and Stanton, 2002) in the driving system has been labelled as equally hazardous to road safety as the cognitive overload that drivers experience when automation fails (e.g. Hancock and Parasuraman, 1992). It would seem that the passive role of the driver to monitor a system of combined Longitudinal and Lateral Control will be less satisfactory than the active role that drivers assume in manual control (Bainbridge, 1983; Stanton and Marsden, 1996). An optimal level of automation, through means of a Driver-Initiated automatic overtake system extension, may be more satisfactory to drivers because they are encouraged to interact with the vehicle more frequently in highway driving. For instance, it is very likely that at some point during highway driving, the vehicle, whether in manual or automated mode, will encounter traffic in the roadway ahead. The first response of the automated system would be to maintain the preset gap, determined by the driver, meaning that vehicle speed may decrease or increase. A 2020 version of the system of Driver-Initiated Automation (Figure 7.3) would begin to monitor the adjacent lanes for further traffic. If a gap is detected, making it possible to perform a lane change manoeuvre safely into an adjacent lane, information will be exchanged between the subcomponents of the automated system and a lane change suggestion will be presented to drivers via the HMI. Drivers can choose to ignore lane change offerings or complete the manoeuvre independently, but if they choose to accept the automated suggestion, the Lateral Controller will deviate from its current lane into a new position in the adjacent lane. At this point, the Longitudinal Controller will then seek to resume the preset speed. Similarly, when the host vehicle has passed the slower vehicle, the system will go through the same processes to offer a return back into the previous lane.

STEP 3: SEQUENCE DIAGRAM AND QUANTITATIVE ANALYSIS

Using the above description, a sequence diagram has been developed to show the interaction that takes place between the driver and other non-human agents within the system network, combining the two stage activation with an automated Lane Change Suggestion (Figure 7.4). This representation shows that much of the additional information that is added into the driving task as a result of automation implementation remains firmly embedded within the automated system architecture, with only the most relevant information being shared with the driver via the HMI regarding system status. On the basis of this interpretation, it would appear that automated system components become central to the functionality of the driving system as the driver delegates increasing levels of control to them.

However, with the emphasis of Driver-Initiated Automation aiming to keep the driver in-the-control-loop, we need to take a closer look at how network dynamism changes as the number of driver commands increases. The first step in addressing how network dynamism is affected by different driver commands is to construct social network diagrams (see Figure 7.5). Communication within social networks

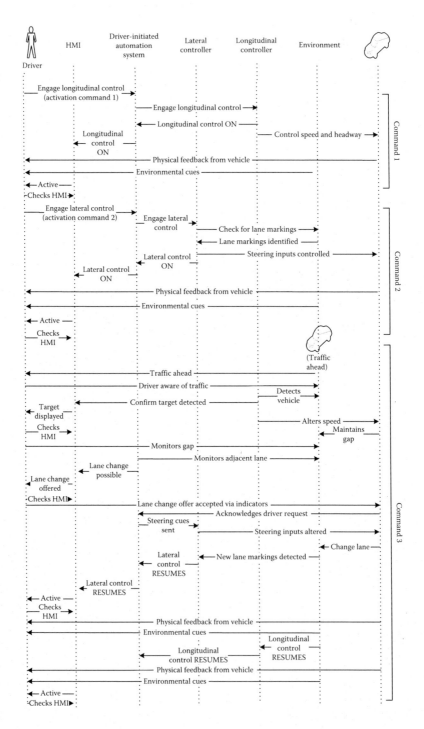

FIGURE 7.4 Interaction between key system agents involved in a driver-initiated automated driving system.

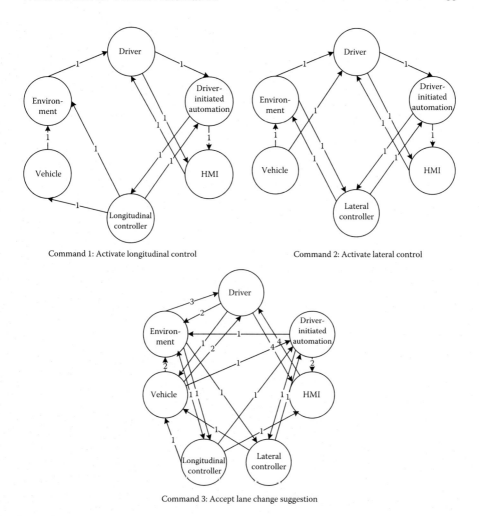

Command 1: Activate longitudinal control

Command 2: Activate lateral control

Command 3: Accept lane change suggestion

FIGURE 7.5 Social network diagrams relating to different driver commands within a driver-initiated command and control system.

is represented by directional arrows and frequencies (Houghton et al., 2006) based upon the interaction patterns presented in Figure 7.4 for each driver command. These social network diagrams represent how key system agents are connected. In this way, it is possible to identify where the system network is most vulnerable, based upon the interaction that takes place between the different system agents (Griffin et al., 2010; Stanton et al., 2015). It is apparent that the most complex network results from the driver command relating to the Acceptance of a Lane Change Suggestion. This is a reflection of the increased communication that is needed between different system agents in fulfilling the command and can be confirmed by quantitative metrics (Houghton et al., 2006) that have been applied to driving (e.g. Salmon et al., 2009; Walker et al., 2011).

TABLE 7.1

Basic Description of Networks

Driver Command	Number of Agents	Number of Edges
Activate longitudinal control	6	10
Activate lateral control	6	12
Accept lane change suggestion	7	19

Further analysis using Agna™ (version 2.1.1; Benta, 2005) was used to examine network density, diameter, cohesion and sociometric status for each of the different driver commands (see Chapter 5 for full definitions and formulae). Table 7.1 provides an overview of the networks and indicates that both driver activation demands (e.g. Activate Longitudinal Control and Activate Lateral Control) can be described as binary (i.e. it can be represented by a zero–one matrix) and non-symmetric (i.e. directed) while the third driver command (Accept Lane Change Suggestion) can be described as a weighted (i.e. non-uniform) and non-symmetric (i.e. directed). The omission of the Lateral Controller agent in the 'Activate Longitudinal Control' social network and the Longitudinal Controller agent in the 'Activate Lateral Control' social network simply reflects the redundancy of each agent in these alternative driver commands.

The level of interconnectedness between individual agents within the system network is represented as a value between 0 and 1 and is a reflection of network density (e.g. Salmon et al., 2014). A value of 1 reflects a fully connected network (Walker et al., 2011) while a value of 0 represents a disconnected network. Table 7.2 confirms that the most densely connected network is the 'Accept Lane Change Suggestion' command, which is also clearly visible in Figure 7.5 with a greater level of communication existing between all system components as task complexity increases. In its current state, the totality of the system network can be described as a medium distributed network that should have some resilience against a network failure.

The fluidity of the network (i.e. the number of 'hops' required to get from one side of the network to the other) is represented by values of network diameter (Stanton, 2014a,b). The results of the analysis are shown in Table 7.3 and indicate that the shortest network is also the most complex (Acceptance of Lane Change Suggestion). During the early stages of automation activation, the system agents work largely

TABLE 7.2

Contrasting Network Density for Different Driver Commands

Driver Command	Density
Activate longitudinal control	0.33
Activate lateral control	0.40
Accept lane change suggestion	0.45

TABLE 7.3
Contrasting Network Diameter for Different Driver Commands

Driver Command	Diameter
Activate longitudinal control	5
Activate lateral control	4
Accept lane change suggestion	3

independently from one another as driving tasks are partitioned gradually. This means that there are less reciprocal links between system agents, as evidenced by Table 7.4, which presents the results for network cohesion. When the system is fully active (i.e. the system is capable of suggesting and executing an automatic lane change), the system agents work cooperatively with one another.

With network cohesion being a measure of reciprocal connections between system agents, Table 7.4 shows that as the driver relinquishes their control over the driving task, the level of network cohesion increases. This simply reflects that the communication between system components increases as the driver is removed further from the control loop. However, just because the driver is removing themselves from active control, it does not mean that they are removing themselves completely from the task. This is because a Driver-Initiated system of automation will continue to function only if the driver continues to issue commands for the automation to complete.

Sociometric status is a useful metric to determine agent prominence within a system network (Houghton et al., 2006). Table 7.5 shows that the intentions of a Driver-Initiated Automation design approach are met, with the driver remaining a key agent within the system network at all stages of the task (i.e. from the initial activation demands to the acceptance of a lane change suggestion). Table 7.5 also highlights other key system agents for different driver commands and indicates that agent prominence within social networks is directly affected by individual driver commands. For example, the driver and lateral controller are the most prominent agents within the social network relating to driver command 2 while the driver, HMI and environment are the most prominent agents within the social network relating to driver command 3. Importantly, as the driving task becomes more autonomous and

TABLE 7.4
Contrasting Network Cohesion for Different Driver Commands

Driver Command	Cohesion
Activate longitudinal control	0.13
Activate lateral control	0.20
Accept lane change suggestion	0.24

TABLE 7.5

Contrasting Sociometric Status for Different Driver Commands

	Command 1 Engage Longitudinal Control	Command 2 Engage Lateral Control	Command 3 Accept Lane Change
Driver	1.00[a]	1.00[a]	2.67[a]
HMI	0.60	0.60	1.83[a]
Driver-initiated Automation	0.80[a]	0.80	1.17
Lateral controller	N/A	1.00[a]	0.67
Longitudinal controller	0.60	N/A	0.83
Environment	0.40	0.80	1.83[a]
Vehicle	0.60	0.60	1.33
Mean	*0.67*	*0.80*	*1.48*

[a] Denotes key system agents based upon the rule that any value above the mean sociometric status value reflects dominance. (Adapted from Salmon, P M, N A Stanton, G H Walker, and D P Jenkins. 2009. Distributed situation awareness: Advances in theory, measurement and application to teamwork. Aldershot: Ashgate.).

the vehicle is controlled automatically by the automated systems (i.e. in the case of command 3), the HMI becomes increasingly important in maintaining and supporting the link between the driver and other system components. This is because it is the only tool that designers can use to ensure that the driver understands 'what' the system is doing at any point in time other than the information that is available from the environment and vehicle.

DISCUSSION

The task analysis of Driver-Initiated Automation may be criticised for being idealistic given its immature development, but it offers a first attempt at describing how information may flow between key system agents during highly automated driving. This chapter shows how the application of EAST can be used to drive the way that automation can be designed to retain the role of the driver within the control-feedback loops. The application of network analysis metrics has revealed that system functionality and resilience to network failure is based upon the 'connectedness' of system agents in allowing the vehicle to perform complex subtasks of driving autonomously. In its current state, the processes that underpin highly automated driving, as presented in this chapter, do appear to maintain the driver in the control-loop despite the delegation of some driving functions being handed to the automation. This is important because the strategy assessed in this chapter was to leave the driver in charge of high-level decision-making, giving permission to the automated subsystems to carry out manoeuvres. The use of network metrics to examine system performance and the role of human agents is becoming more popular (Stanton, 2014a; Stanton et al., 2015).

Even so, the consequence of this control transfer on driver behaviour remains unknown, and more research is needed to ensure that the new driver role afforded by Driver-Initiated Automation is appropriate. The representations presented in this chapter offer a visualisation of how an 'ideal' network may function. Realistically, however, prolonged periods of driver inactivity (i.e. 'hands and feet free' driving) could result in issues surrounding driver disengagement, boredom and fatigue (e.g. Stanton et al., 1997; Young and Stanton, 2002). In other words, highly automated driving (such as a system that automates longitudinal, lateral and overtake manoeuvres) is likely to divert the driver's attention away from the road to other tasks (de Winter et al., 2014). Merat et al. (2014) found that both the level and the type of automation can have a direct effect on levels of driver engagement. However, while the likelihood of drivers engaging in non-driving tasks increased as the level of automation increased, this was not detrimental to driving in typical conditions. In addition, Jamson et al. (2013) found that when drivers experienced highly automated driving, they were less inclined to change lanes even in heavy traffic situations despite increased journey times. This suggests that a Driver-Initiated automatic overtake may not protect against driver disengagement in the way that is hoped, but further research is needed to validate these findings. Of greatest concern is the need for drivers to resume control in atypical driving situations, which could result in sudden changes to driver workload (e.g. Stanton et al., 1997; Jamson et al., 2013) that could be detrimental to driving safety (Rudin-Brown and Parker, 2004). This is because increased eyes-off-road time (e.g. Peng et al., 2013) is associated with reduced driver SA (Dozza, 2012; Young et al., 2012). One of the real concerns for highly automated driving systems of the future is how driver workload can be managed so that situations of mental underload and overload do not develop (Stanton et al., 1997; Young and Stanton, 2002, 2004).

In addition, driver familiarity with system operation is also likely to influence driver usage patterns. For example, there are a number of common events that automated longitudinal control systems (such as ACC) are unable to cope with, which means that the driver needs to resume control. These functional limitations are discovered and learned as a result of experience with the system (e.g. Larsson, 2012; Larsson et al., 2014; Strand et al., 2011). As drivers begin to learn the functional limitations of the automated system, they begin to build an understanding of what the system is capable of (Rasmussen et al., 1994). If the system is highly reliable, driver expectations are continually reinforced, which may make them more susceptible to instances of 'automation surprise' or 'startle' in the case where the automated system behaves in an otherwise unfamiliar way (Sarter et al., 1997). Negative experiences such as this have been found to affect subjective ratings of trust (e.g. Wiegmann et al., 2001), which may lead to disuse (drivers reject the benefits of the system), misuse (drivers become complacent) or rejection (drivers will not use the system even when available) (e.g. Parasuraman et al., 1993; Sheridan, 1988). In addition, increased reaction times (e.g. Merat and Jamson, 2009; Young and Stanton, 2007b) are thought to be associated with issues of cognitive underload (Vollrath et al., 2011), overload (Stanton et al., 2011) and reduced responsibility (e.g. Farrell and Lewandowsky, 2000); all of which signal that driver desensitisation is a real concern in automated driving and should not be ignored.

More research is therefore needed to establish whether a command and control relationship between the driver and the automation is sufficient to keep drivers actively engaged in the task (Stanton and Baber, 2006) and reduce the risk of driver disengagement. The driver should be required, or at the very least encouraged, to interact with the vehicle throughout the journey. Either way, it seems increasingly important for automated systems to be aware of the driver's state and have the ability to re-engage if desensitisation does occur (Merat et al., 2014), signalling further challenges in the quest to reach higher levels of driving autonomy. We may find that while a reliable system that will not breach driver expectations can be achieved, it may not reflect the highest capabilities of the technological components involved. Instead, we need to decide whether the most 'capable' system should be balanced with the most 'user-friendly system' (Zheng and McDonald, 2005). Future research should test the hypotheses developed in this chapter to empirically validate the findings. Using a highly instrumented vehicle capable of automating longitudinal, lateral and overtake manoeuvres will substantially increase our knowledge of driver behaviour and their interaction with both the vehicle and the environment. Valero-Mora et al. (2013) claim that the use of such vehicles in a relatively naturalistic environment can significantly contribute to driver–automation interaction research that can overcome some of the issues associated with simulation.

FUTURE DIRECTIONS

Chapters 8 and 9 continue to extend our understanding of the role of the driver within a highly automated driving system that adopts Driver-Initiated design through experimental investigation on multi-lane carriageways in live traffic. Going beyond driving simulation gives rise to unique insights into the changing nature of driver–vehicle interactions in a real-world setting.

8 Discovering Driver–Vehicle Coordination Problems in Early-Stage System Development

INTRODUCTION

A critical question being asked by many vehicle manufacturers is what actually happens when the driver finds themselves being 'hands and feet free' within their vehicles. This case study was used as an investigation into the functionality of Driver-Initiated Automation (both system capabilities and architectural issues) at a very early phase of system development. Using a selection of popular Human Factors tools, a multidisciplinary team of researchers, engineers and systems designers wanted to explore how the use of a highly instrumented vehicle could be used in extending our understanding of driver–vehicle interaction patterns under high levels of driving automation. The use of a highly instrumented vehicle offers a step forward from traditional driving simulator studies (Valero-Mora et al., 2013) as research can be carried out in a more naturalistic driving environment.

With this in mind, the purpose of this chapter was to essentially validate and further explore the interaction occurring between the driver and the automated system of Driver-Initiated Automation that was revealed in Chapter 7. Thus, in terms of the Systems Design Framework, this chapter is primarily focussed upon the link between Phase 1 and Phase 2, as shown in Figure 8.1. The hope was to extend our understanding of Distributed Cognition (Hutchins, 1995a) in highly automated driving systems.

METHOD

The work presented in this chapter is based upon an exploratory study by Banks and Stanton (2015b).

Participants

Two participants with Advanced Driver Training were recruited to take part in this study due to the exploratory nature of a highly automated prototype technology. One participant was an experienced user of Driver-Initiated Automation having built up a

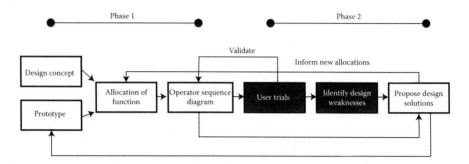

FIGURE 8.1 Aspects of the Systems Design Framework focussed on this chapter as highlighted in grey.

number of hours using the system the preceding week while one participant was an inexperienced, first-time, user of the system.

Ethical permission to conduct the study was sought and granted by the sponsoring company.

Experimental Design and Procedure

After gaining informed consent, both drivers were instructed to complete a much condensed version of the Dundee Stress State Questionnaire (DSSQ; Matthews et al., 2002) that included the Energetic Arousal and Tense Arousal subscales. According to Matthews et al. (2013), Energetic Arousal can be seen as similar to the level of task engagement while Tense Arousal can infer levels of distress. The two subscales consisted of 39 items that were coded as required by the developers (Matthews et al., 2002). In addition, the inexperienced user was also given a brief introduction to the functionality of the Driver-Initiated system by means of images of the HMI. This was so that the inexperienced user could familiarise himself with the functioning of messages and controls. As new vehicle consumers do not receive any training in the use of vehicle subsystems, a description like this was considered to be consistent with current custom and practice. Upon completion of the pre-drive DSSQ, drivers were able to familiarise themselves with the host vehicle as they made their way from the University of Warwick (United Kingdom) campus to the A46 dual carriageway, a distance of approximately 2 miles. The host vehicle was a medium-sized family saloon equipped with both radar and Light Detection and Ranging (LIDAR) sensors that analysed the surrounding environment and monitored for other road obstacles and lane markings.

The test route consisted of a 16-mile stretch of the A46 between Coventry and Warwick that took approximately 20 min to drive. Throughout this time, drivers were invited to complete a verbal commentary recorded using Smart Voice Recorder version 1.7. Upon joining the A46, drivers were invited to use the Driver-Initiated automated system given the understanding that drivers would manually override the subsystem when necessary (e.g. in the case of malfunction which caused automation to drop out). A Safety Driver sat in the passenger seat and could answer any questions that the driver posed. An experimenter sat in the rear of the vehicle. Upon return to the University of Warwick campus, drivers were instructed to complete the post-drive DSSQ and the National Aeronautics and Space Administration Task Load Index (NASA-TLX; Hart and Staveland, 1988) to assess workload during the task.

Data Reduction and Analysis

Once verbal commentaries had been transcribed, an initial coding scheme based upon systemic SA research (e.g. Walker et al., 2011) was used to analyse the content of verbal reports. Refinement of this coding scheme ensued using a hybrid of theory-driven and data-driven approaches. The iteration process continued until the verbal reports were judged to be adequately categorised into the coding scheme. The final coding scheme consisted of seven categories. Table 8.1 presents this coding scheme along with descriptions and examples.

RESULTS

Thematic Analysis

Figure 8.1 shows the total number of observations made for both the experienced and inexperienced system users. Unsurprisingly, the inexperienced user generated evidence of a greater number of functionality concerns characterised by an increase in questions posed to the Safety Driver to seek validation on system behaviour. These questions typically focussed on clarification of system behaviour, the meaning of HMI content and system limitations (Figure 8.2).

Table 8.2 shows the code frequency as a percentage of total coding and indicates that the inexperienced user of the system was heavily focussed upon functionality issues and building their knowledge database of system functionality while the experienced user was more evenly spread.

TABLE 8.1
Coding Scheme for Verbal Commentaries Including Descriptions and Examples

Code	Description	Example
Behavioural disparity	Disparity in system performance and what the driver would normally do	'See really I would have pulled over by now'
Driver knowledge	Reference to driver knowledge of system behaviour/operation	'This wouldn't let me do that'
Other traffic	Any reference to the behaviour of other traffic	'You can never really second guess what other people are going to do'
Driver behaviour	Statements referring to own behaviour	'Quite happy to take my hands off the wheel'
Manual override	Evidence of the driver regaining control of the vehicle	'I'll just do it manually'
System behaviour	Overt references to system operation	'It's keeping me in the lane'
Functionality issues	A lack of understanding surrounding system function	'So it's still working now?'

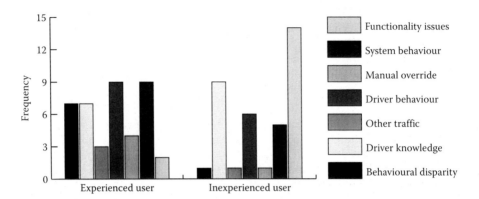

FIGURE 8.2 Frequency of code occurrence following thematic analysis of driver verbalisations.

Evidence of Driver–Vehicle Coordination Problems

Interestingly, the verbal commentary provided some evidence of mode confusion on behalf of the inexperienced user:

> Thanks for telling me because I didn't spot the lines [on the HMI display]. So because I'm unsure of what state it's in, what I'm going to do is press the brake and I'm going to start all over again... [inexperienced user]

Mode confusion occurs when the human operator of a system fails to understand the current and future state or behaviour of automated subsystems (see Sarter and Woods, 1995; Stanton and Salmon, 2009). In this case, the driver believed that the system was on when actually it was not (e.g. Sarter, 2008). This statement implicates the importance of HMI design and suggests that the current prototype lacked transparency (Stanton and Marsden, 1996) although these types of error might reduce over time as experience in using the system increases:

TABLE 8.2
Code Frequency as a Percentage (%) by User

Code	User	
	Experienced	Inexperienced
Behavioural disparity	17	3
Driver knowledge	17	24
Other traffic	7	3
Driver behaviour	22	16
Manual override	10	3
System behaviour	22	13
Functionality issues	5	38

I think it's quite clear and precise really. You know exactly what you're being offered and when [experienced user]

The experienced user of the system provided evidence of a good working knowledge of the vehicle system and appeared at ease throughout the drive apart from when the automation behaved in a way that was unexpected (e.g. automation surprise; Sarter et al., 1997). This behaviour deviated from their established mental model of system operation signalling a breakdown of driver–automation coordination. At this point, the driver sought clarification from the Safety Driver and appeared anxious:

'What happened there? … I'm just a bit more aware, there's a few things that happened back there that makes me definitely keep in control of it'

This unanticipated system behaviour challenged internal mental models surrounding system functionality and disrupted normal data-driven and knowledge-driven monitoring of the system (Sarter et al., 2007). It is errors like these that have the potential to result in future accidents, especially if the automation behaves consistently for prolonged periods enabling drivers to become complacent. Complacency may have happened to the experienced user of the system:

'I was thinking it's going to be a breeze on the way down, not a problem and then it went and did something like…. But like you say, on the way back I felt a lot more comfortable. It's knowing exactly what it's going do and what it's capable of'

Subjective Stress and Workload

The results of the DSSQ are presented in Figure 8.3 and show a shift in Energetic Arousal and Tense Arousal by both users. The post-drive scores indicate that the experienced user of Driver-Initiated Automation became less energetically aroused (engaged in the task) and more tensely aroused (stressed by driving) while the inexperienced user became more energetically aroused and less tensely aroused. Desmond and Matthews (2009) reported that prolonged driving can produce a loss of task engagement and this appears to be true for the inexperienced user of Driver-Initiated Automation. The control transition that took place between the inexperienced user and the automated system appeared to lower task demand and subsequent stress levels.

As Hockey (1997) proposed that the degree of effort required to sustain system performance is directly related to the level of task demand, it comes as no surprise that the inexperienced user reported lower scores on all but one dimension of the NASA-TLX (Figure 8.3). These findings are very similar to those found by Young and Stanton (2002). The experienced user experienced increased workload, perhaps attributable to having greater knowledge of system functionality and its subsequent limitations. This may have led to an increase in the experienced user monitoring behaviour, perhaps as a result of the behavioural disparity that was experienced between the experienced user and the automated system (as evidenced by the VPA). In addition, the experienced user was potentially more capable of knowing when the automated system was behaving unusually (Figure 8.4).

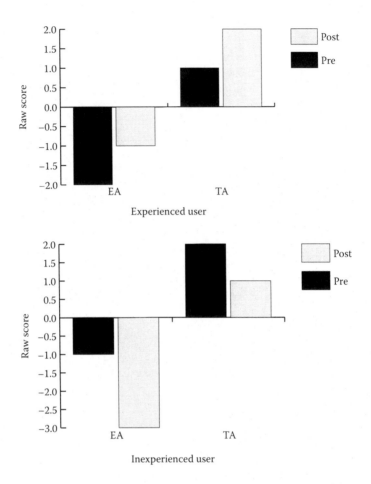

FIGURE 8.3 Comparison of scores on the DSSQ for Energetic Arousal (EA) and Tense Arousal (TA) for the experienced and inexperienced users.

An automation surprise is likely to be more stressful than a mode confusion because it challenges pre-existing mental models (e.g. Hoc et al., 2009; Revell and Stanton, 2014) whereas the mode confusion is most likely to occur when mental models are still being constructed. It has been previously suggested that the workload imposed by a task can have a direct effect on objective performance and subjective stress response (Matthews et al., 2002). These results support this claim.

DISCUSSION

Although the use of verbal reports is highly debated (e.g. Baumeister et al., 2007; Boren and Ramey 2000; Ericsson, 2002; Jack and Roepstorff, 2002; McIlroy et al., 2012), they offer a means to explore the momentary thoughts related to driver–vehicle coordination problems. This includes effects of sudden demand transition, which can be supported through the use of subjective measures (e.g. Helton et al., 2004).

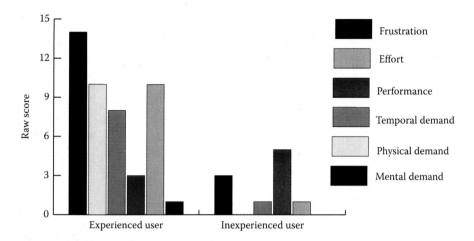

FIGURE 8.4 Subjective workload scores across the six dimensions of the NASA-TLX for both the experienced and inexperienced system users.

Without verbal commentary, it seems unlikely that the problems experienced by the drivers would have been captured. Observation alone did not reveal any problems with driver–vehicle coordination. It was only with the analysis of the verbal commentary that these coordination problems became apparent.

Much like in aviation, automation surprises within driving are likely to be experienced by all drivers, regardless of their experience in using the system. However, it is more likely to cause greater stress to those with greater usage (Sarter et al., 1997). This is because experienced users may have created more robust mental models about how the system functions and have developed more trust in the system (due to its perceived reliability). In contrast, new users of a system remain flexible to change as new experiences that would otherwise be perceived as 'unexpected' help create these robust models in the first place.

Stanton and Marsden (1996) proposed that one of the issues to the introduction of automated aids is when the system itself fails to deliver the expected benefits outlined during its design. These performance shortfalls on the part of the system may actually lead to an increase in accidents due to confusion over system state or not behaving in the way expected by the driver. As integrated Advanced Driver Assistance System (ADAS) becomes more common on the road to full vehicle automation, it seems likely that the prevalence of automation surprise in a driving environment will become more common. This is because automation remains incapable of coping with all driving eventualities (Norman, 1990). While the prevalence of mode confusion may reduce as drivers quickly learn the different system states (Larsson, 2012), it would seem that a brief introduction to Driver-Initiated Automation was not sufficient enough to avoid mode confusion in this study. Further investigation is required to see whether the mode confusion reported in this study was in fact a designer error (e.g. Chapanis, 1995).

The results of this study highlight the importance of maintaining drivers in-the-loop to ensure they remain sensitive to changes within their environment especially during the intermediate levels of automation (Endsley, 2006; Flemisch et al., 2012). If control is transferred back to drivers when they are least expecting it, their ability to take back control may be restricted and system performance will be significantly affected. Encouragingly, the results of this study demonstrate that far from being removed from the control-feedback loop (Stanton et al., 1997, 2007b), the setup of Driver-Initiated Automation maintained pre-automation driver status, meaning that driver–vehicle coordination problems were quickly and effectively overcome. The irony of automation as discussed by Bainbridge (1983) is that highly automated systems still require human operators as automated systems have restricted functional envelopes (Zheng and McDonald, 2005). However, as long as automation remains 'adaptable', the division of labour between the driver and automated systems can remain dynamic and flexible (Parasuraman, 2000). This means that any deviation from normal system behaviour can be quickly addressed through a swift control transfer back to the driver as was the case of the driver–vehicle coordination problems found in this study.

It seems likely that the greatest obstacle to overcome in terms of driver–vehicle coordination problems is issues surrounding driver complacency. It is evident that keeping the driver in-the-loop does not safeguard against this phenomenon. Continued research is needed to ensure that overall system safety can be maintained after prolonged periods of reliable automation.

PRACTICAL RECOMMENDATIONS FOR FUTURE RESEARCH

The research conducted by Banks and Stanton (2015b) offers a very unique opportunity to observe driver–vehicle coordination problems in a more naturalistic setting than that of driving simulator studies. However, there were a number of practical constraints that limited the feasibility of data collection that were beyond the control of the researcher. These included the following:

- *Commercial Sensitivities*: The Driver-Initiated feature used in this study was an early prototype model. At this stage of development, it was not possible to share detailed information relating to any of the vehicle technology on-board. Due to these sensitivities, this study was constrained to using individuals who held an up-to-date non-disclosure agreement with the sponsoring company and external supplier. Although this issue is not easily overcome in the short term, future research should attempt to widen the demographic of the participant pool to include individuals with a non-engineering background.
- *Test Vehicle Availability*: The test vehicle was only made available for a limited testing period. Although the experimental design had been planned in advance, the testing schedule was significantly shortened due to other commercial needs. This meant that only two drivers could be selected to take part in the study. It is highly recommended that future research should guarantee access to the vehicle to ensure a greater number of participants can take part. An official testing period or 'User Trial' would ensure that experimental design and procedures can be followed as planned.

- *Legal Constraints*: For insurance purposes, only drivers with Advanced Driver Training and who were employees of the sponsoring company could take part. This was a mandatory requirement for any research being conducted with the use of a prototype technology. However, this was also likely to bias the data to some extent as the sample used in this study was unlikely to be representative of a typical driver population. As employees, they also held a vested interest in the success of future market deployment. Although these legal constraints are difficult to overcome in the short term, the use of a Safety Driver in future research may mean that individuals without Advanced Driver Training will be able to participate. Given appropriate permissions, participants representing a typical driver population may be recruited.
- *Sample Size and Demographics*: As an exploratory investigation into driver–vehicle–world coordination problems, this study proved to be an invaluable source of information for system designers at the sponsoring company. However, future research should make use of a larger sample size of mixed age, gender and experience in the use of reliable Driver-Initiated Automation. Even so, the opportunity to observe driver behaviour in a naturalistic 'hands and feet free' driving system was worthwhile, especially when considering that there has been growing concern about what drivers may do if they are not in active control. In addition, with concerns growing over how well drivers will cope in the event of system failure (e.g. Hoc et al., 2009; Shorrock and Straeter, 2006), this research provided some encouraging results. Although it seems unlikely that the final Driver-Initiated Automation product would elicit the same sort of automation surprise observed in this study, both drivers were quickly and efficiently able to regain control of the vehicle despite a sudden increase in subjective stress levels.

FUTURE DIRECTIONS

Chapter 9 continues to build upon this chapter by further exploring the relationship between the driver and the automated prototype in a real-world setting. Chapter 9 specifically seeks to address issues relating to HMI design given the coordination problems that were revealed in this chapter.

9 Driver-Initiated Design
An Approach to Keeping the Driver in Control?

INTRODUCTION

We learnt from Chapter 8 that driver expectations of system operation can heavily influence the way in which drivers perceive a 'new' driving feature. The Technology Acceptance Model (Davis et al., 1989) postulates that while intentions to use a technology affect subsequent usage behaviour, the perceived ease of use (i.e. systems usability) is also likely to determine the intention to use. If, for example, drivers perceive an automated system to be confusing, they are less likely to use it. Previous literature has found that driver attitudes are more positive when automated assistance is available during otherwise monotonous driving situations (e.g. Fancher et al., 1998). For example, automated assistance on highways brings the added benefit of improving driver comfort and convenience (e.g. Saad and Villame, 1996). It is important therefore to explore the appropriateness and acceptability of a Driver-Initiated system of automation on subjective reports of driver stress and workload.

USABILITY OF DRIVER-INITIATED AUTOMATION

The primary purpose of this study conducted by Banks and Stanton (2015c) was to conduct an initial assessment of a prototype system of Driver-Initiated Automation combining lateral and longitudinal control. In addition, an automatic overtake system was also developed that could either be accepted or ignored. Such a system was described in Chapter 7. The main purpose of which was to assess the systems design effects on subjective reports of driver mental workload and trust as well as to gain some insight into the design of the HMI. The purpose of the latter assessment was so that any potential design weaknesses within the prototype HMI architecture could be highlighted given the coordination problems highlighted in Chapter 8. In addition, results of any subsequent analysis would provide recommendations for suitable revision that would improve system transparency. This is because it was recognised that the 'weakest link' within a Driver-Initiated system could lie between the HMI and the driver, which was revealed in Chapter 7 and by initial investigations reported in Chapter 8. This process therefore reflects a continuation of Phase 2 research using the Systems Design Framework shown in Figure 9.1.

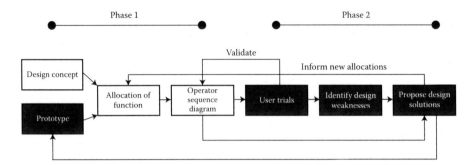

FIGURE 9.1 Aspects of the Systems Design Framework focussed on this chapter as high-lighted in grey.

METHOD

Participants

A total of 32 participants (mean age = 38, SD = 10.8) were recruited from the staff cohort of the sponsoring company. The study was made available to all employees via an internal Google Calendar where they were asked to select a time and date most suitable to them. All participants held a full U.K. driving licence and were between the ages of 25 and 60. This was to ensure that the performance decrements often dem-onstrated by older drivers (i.e. over 65s) did not affect the results of the study.

Ethical permission to conduct the study was sought and granted by the sponsor-ing company. A comprehensive risk assessment included a number of mitigation measures. For example, the role of the Safety Driver was not only to monitor the roadway environment to ensure that manoeuvres could be performed safely but also to ensure that if for any reason drivers failed to regain control of the vehicle follow-ing an audible and visual system warning, they could provide verbal instruction. Importantly, all participants were told that while system warnings may occur in the automated driving condition, a total loss or failure of automation would be unlikely to occur. Even so, the Safety Driver could override the automation completely (i.e. switch back to manual) by pressing a button if needed. In addition, a Safety Vehicle was used to monitor the traffic ahead of the test vehicle and communicate any haz-ardous situations, such as harsh braking, to the Safety Driver via radio link.

Experimental Design and Procedure

The test vehicle was a left-hand drive, medium-sized family saloon car equipped with a prototype system of longitudinal and lateral control that allowed for the auto-mation of driver-initiated overtake manoeuvres (i.e. pull out and pass). Although participants were U.K. licence holders, they were all familiar with driving left-hand vehicles on U.K. roads, which was an essential demographic criterion to ensure that task complexity was not inadvertently increased by a lack of experience in driving left-hand vehicles.

Upon providing informed consent, participants were given an introduction to the functionality of the system within the vehicle. Within this introduction, drivers were presented with a series of icons that they may see on the HMI. The HMI was

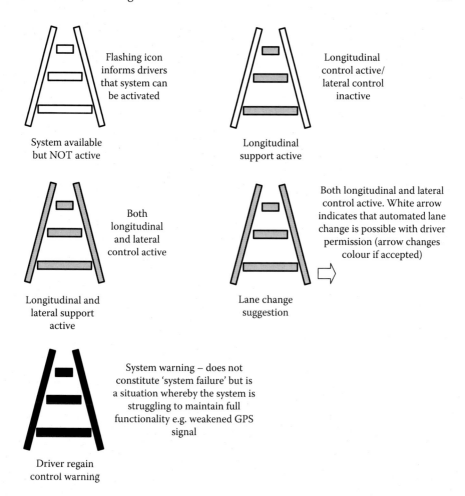

FIGURE 9.2 Schematic representation of display icons located on head-up HMI.

represented on a Head-Up Display situated on the windscreen. A schematic representation of this interface is shown in Figure 9.2. Following this introduction, drivers were invited to familiarise themselves with the controls and ask any questions.

Drivers were requested to complete two driving conditions within a 20.4-mile circular test route along the M40 motorway between Junctions 12 and 15: manual and automated. The presentation of these conditions to drivers was counterbalanced to eliminate order effects. Upon joining the carriageway, drivers were instructed to maintain a speed of 110 kph (70 mph) and abide by U.K. driving law at all times. Drivers were not directly invited to drive 'hands free' at any time but were invited to drive in a manner they felt comfortable with. Throughout each experimental condition, drivers were invited to perform three basic driving manoeuvres: maintain speed and distance to Target Vehicle, perform an overtake (pull in and out) and finally perform a lane change without the use of directional indicators when safe to do so. In the manual driving condition, the autonomous feature was not active. In the automated driving condition, drivers were invited to activate the automated system as soon as they had safely joined the highway. Drivers were prompted to perform these manoeuvres by a Safety Driver who sat in the front passenger seat and upon completion were asked a series of questions by the researcher who sat in the back of the vehicle. These questions were specifically designed to assess driver knowledge and understanding of the HMI prototype in an effort to highlight design weaknesses within the current system architecture. Responses were recorded through the use of video and audio equipment as well as written observational notes. The content of the interview was subjected to thematic analysis as has been previously demonstrated by Banks et al. (2014a,b). In addition, drivers were invited to complete the NASA-TLX (Hart and Staveland, 1988) and Checklist for Trust between People and Automation (Jian et al., 2000), following each driving condition to measure subjective ratings of workload and trust.

Data Reduction and Analysis of Observational Data

An initial coding scheme was developed using a data-driven approach. The aim of the coding scheme was to reveal information relating to system usability and driver behaviour. A small focus group consisting of two design engineers and two Human Factors researchers read through all of the driver responses to the structured questionnaire and began to highlight key themes. Following repeated iterations, the focus group agreed on a final coding scheme consisting of five key themes. The final coding scheme with definitions and descriptions is shown in Table 9.1. Four verbal transcripts (representing approximately 10% of the sample) were selected at random to be subjected to further analysis by a secondary coder to calculate inter-rater reliability. This analysis resulted in 97.88% agreement, which is over the 80% threshold (Marques and McCall, 2005).

RESULTS

Thematic Analysis

Results were analysed based upon the frequency of code occurrence in each driving condition. The results are shown in Table 9.1. They provide a useful, albeit, exploratory insight into driver perception of system usability and possible design deficiencies within the current prototype architecture. Results of individual key themes are discussed in turn.

Knowledge of System Engagement

Figure 9.3 shows the frequency of occurrence for the Knowledge of System Engagement key theme. It indicates that the primary information used by drivers to ascertain if the system was engaged or not was the Head-Up HMI display for both manual and automated driving. This suggests that the interaction or link that exists between the driver and the HMI agent is extremely important. Figure 9.4 also indicates that the emphasis that drivers placed upon each category was affected by the level of assistance that they received. For example, drivers were more likely to reference being in control and using information from their environment to aid their understanding of system engagement in manual driving, while more emphasis was placed upon physical feedback from the vehicle in automated driving.

TABLE 9.1

Coding Scheme, Descriptions and Frequency of Occurrence for Manual Driving and Automated Driving While the System of Driver-Initiated Command and Control Was Active

Key Theme	Subthemes	Description	Frequency	
			Manual	Automated
Knowledge of system engagement	Driver in control	References to the driver being in control/not having engaged system	43	17
	Information from the environment	References to look at other traffic, lane markings	29	15
	HMI display	References to the colour of lines/icons	67	65
	Physical feedback	References to physical feedback from the steering wheel	14	39
	Automation in control	References to the automation being in complete control of vehicle operation	N/A	8
Expectation management	Expectation met	Realistic expectation of system operation	80	76
	Expectations unmet	Unrealistic expectation of system capabilities	15	7
	Unknown expectation	Driver did not know what to expect	2	3
Behavioural observations	Driving 'hands off'	Observation of 'hands free' driving	N/A	41
	Failure to regain control following system warning	Observation that drivers required prompts to regain control of the steering wheel	N/A	7
System usability	Required assistance from safety driver	Verbal instruction given to driver regarding system functionality/meaning	2	11
	Unexpected lane change occurred	Vehicle moved across two lanes instead of one	N/A	3
	System initiation problems	Struggling to turn the system on or off	N/A	9
	Perception of unsafe lane offerings	Mismatch between driver perception of safety and what the system suggests is safe	N/A	2
	Misunderstanding of HMI display	Misinterpretation of HMI display	22	6
	Correct understanding of HMI display	Correct interpretation of HMI display	40	57
System mode	System on	Driver reports system is on	7	76
	System off	Driver reports that system is off	85	11
	Unsure	Driver reports that system status is unknown	4	3

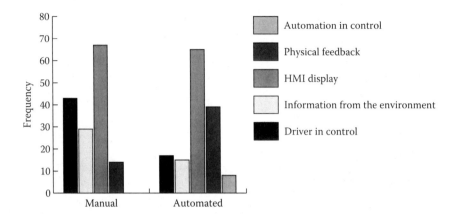

FIGURE 9.3   Frequency of subthemes regarding Knowledge of System Engagement.

Expectation Management

The management of driver expectations regarding their subjective perception of 'how' the system would behave was on the whole very good (Figure 9.4). However, driver expectations of system behaviour were more likely to be unmet in manual driving. Reasons for the mismanagement of driver expectation may be found in an analysis of codes relating to system usability.

System Usability

Figure 9.5 shows that while it is encouraging to note a high number of positive references to understanding the HMI were observed for both manual and automated driving, drivers were more likely to misunderstand the meaning of the HMI in manual driving. A common mistake was for drivers to think the system was actively assisting the driver when it was not, representing a form of mode confusion (Norman, 1990; Sarter and Woods, 1995; Stanton and Salmon, 2009). In many

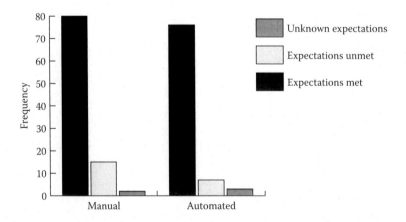

FIGURE 9.4   Graphical representation of subthemes relating to Expectation Management.

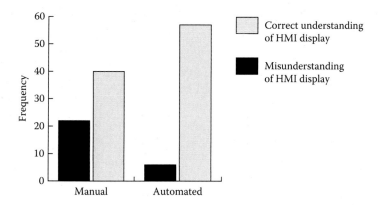

FIGURE 9.5 Driver understanding of HMI content based upon the level of automated assistance.

instances, this represented that driver expectation of system functionality went beyond the original design parameters, hence the increased frequency of driver expectations being unmet, signalling an issue within the feedback presented to the driver via the HMI. Even so, drivers were able to correctly identify the system state most of the time: 89% correct in manual driving and 85% correct in automated driving.

In addition, code frequencies for the remaining subthemes of System Usability highlight a number of other important considerations that may affect the way in which drivers choose to use the automated system in the future (Figure 9.6). For example, unexpected lane changes (caused directly by the driver inaccurately using the controls) and the perception of unsafe lane offerings, although experienced by only a small number of participants, indicate that the Driver-Initiated system in its current state requires some modification. The implications of these 'negative' experiences are likely to be reflected in subjective trust and workload ratings.

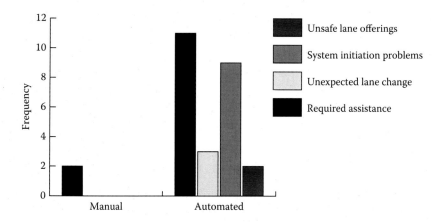

FIGURE 9.6 Frequency of automated system design issues experienced by drivers.

Behavioural Observations

Out of a possible 96 instances where the driver could become 'hands and feet free' while driving in automated mode, 41 observational references, made by the experimenter who sat in the back of the test vehicle, were made to being 'hands free'. This means that almost half of the manoeuvres in the automated driving condition saw the driver adopt a more 'supervisory' role, allowing the vehicle to perform the manoeuvre or task autonomously. This is an important observation for designers to consider, especially given that in the current study, drivers were told to remain in control of the vehicle at all times. Their willingness to allow the vehicle to take full control could signal a form of complacency.

In addition, and what is more concerning, is that seven drivers had to be prompted to regain active control of the vehicle following a system warning. Although this represents a small proportion of the sample, it highlights a severe deficiency in the current design of the warning system.

Driver Trust

Results of the 7-point Checklist for Trust between People and Automation (ranging from 1 'Not at all' to 7 'extremely') are presented in Figure 9.7 and show that driver responses to negatively framed questions are consistently rated less favourably for automated driving in comparison to manual driving. Wilcoxon signed-rank tests revealed the following: deceptive ($z = 2.532$, $p < 0.05$, $r = 0.45$), underhanded ($z = 2.076$, $p < 0.05$, $r = 0.37$), suspicious ($z = 3.749$, $p < 0.01$, $r = 0.66$), wary ($z = 3.306, p < 0.01, r = 0.58$) and harmful ($z = 2.864, p < 0.01, r = 0.51$). This may be attributed to drivers not having yet learnt the competence limits of the technology (Fitzhugh et al., 2011) or having experienced some of the issues highlighted above (e.g. unsafe lane offering). It may also signal underlying issues, such as a refusal to transfer control to an automated system despite Driver-Initiated design that is used to maintain a command–control relationship between the driver and the automation.

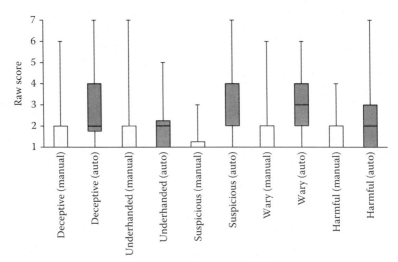

FIGURE 9.7 Responses to negatively framed questions.

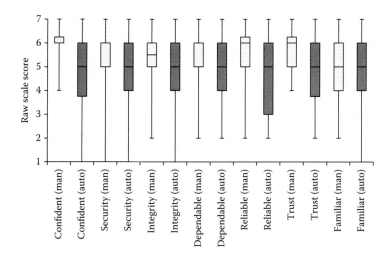

FIGURE 9.8 Responses to positively framed questions.

Again, manual driving was rated more favourably than automated driving when drivers were asked positively framed questions as shown in Figure 9.8. Wilcoxon signed-rank tests revealed the following: confident ($z = -3.546, p < 0.05, r = -0.63$), security ($z = -2.170$, $p < 0.05$, $r = -0.38$), dependable ($z = -2.999$, $p < 0.05$, $r = -0.53$), reliable ($z = -2.974$, $p < 0.05$, $r = -0.53$) and trust ($z = -3.469$, $p < 0.05$, $r = -0.61$). The greater range in driver responses to automated driving is likely to be a reflection of the consequence of first-time system use in addition to the discussion of negatively framed questions. In order to encourage driver trust in automation (Ashleigh and Stanton, 2001; Lee and See, 2004; Stanton et al., 2011), drivers need a clear understanding of what the system is capable of and its purpose (Rasmussen et al., 1994). Any violation of a driver's expectation of system functionality is likely to have an effect on subjective ratings of trust. For example, Dzindolet et al. (2002) propose that naive operators are more likely to expect automated assistance to be capable of outperforming them. If the automation fails to perform in the way expected, ratings of trust begin to decline (Wiegmann et al., 2001). Future research should seek to expose drivers to longer periods of highly automated driving to see if subjective ratings of trust change over time.

A 'negative' first-time experience of using a new automated system is also likely to affect subjective ratings of driver workload as internal mental models are continually challenged as the driver attempts to build a picture of how the automated system works. Any sudden or unexpected system behaviour is likely to induce stress and increase workload. For example, an unexpected lane change could lead to an automation surprise (Sarter et al., 1997) that could result in a sudden increase in driver workload as they attempt to understand 'why' the system is behaving in this way as well as inducing driver stress.

Driver Workload

Analysis of the NASA-TLX revealed that median overall workload scores were significantly higher in automated driving (Mdn = 42) in comparison to the manual driving (Mdn $= 26.5$) ($z = 3.107$, $p < 0.005$, $r = -0.55$). This difference is shown in Figure 9.9.

Further analysis of the individual subscales of the NASA-TLX revealed significant differences between mental demand ($z = 3.327$, $p < 0.005$, $r = -0.59$), temporal demand ($z = 3.134$, $p < 0.005$, $r = -0.55$), effort ($z = 2.409$, $p = 0.05$, $r = -0.43$) and frustration ($z = 2.843$, $p < 0.005$, $r = -0.50$) with automated assistance consistently resulting in increased ratings, as shown in Figure 9.10. This on the one hand may be a simple reflection of the fact that these ratings were collected during first-time use of the automated system. However, it may also signal more important issues that require consideration. For example, increased workload could be a reflection of the additional requirement for drivers to monitor system behaviour and ensure the vehicle was responding effectively, in addition to traditional driver monitoring of other traffic on the road (de Winter et al., 2014; Merat et al., 2012; Stanton and Young, 2005; Stanton et al., 1997, 2001; Young and Stanton, 2002) as they develop their internal working models. This means that although the driver was not in direct control of vehicle outputs, they had to remain aware of changes in their environment (Parasuraman and Wickens, 2008), suggesting that far from reducing workload, automation may simply shift driver attention to other tasks such as system monitoring (Reinartz and Gruppe, 1993). This additional responsibility could be enough to increase subjective workload ratings (Stanton et al., 1997, 2001, 2005). However, over time, subjective workload ratings of driving may begin to decrease as additional attentional resources are released to complete other tasks (Liu and Wickens, 1994; Rudin-Brown and Parker, 2004; Stanton et al., 2001). de Winter et al. (2014) argued that automation of longitudinal and lateral control is distinctly different from traditional ACC because it has the potential to divert driver attention to secondary tasks. In addition, Carsten et al. (2012) report that drivers are more likely to engage in other tasks when they receive lateral support. Future research should expose drivers to increased duration of automated driving to see how comfort levels and ratings of workload change over time. This would be especially useful to see how levels of driver engagement are affected by increased durations of automated control.

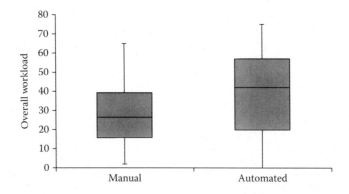

FIGURE 9.9 Overall workload scores of the NASA-TLX between manual and automated driving conditions.

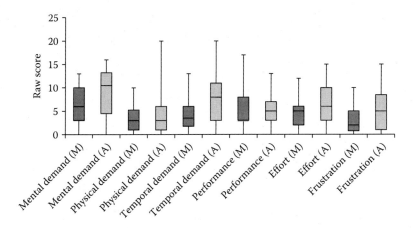

FIGURE 9.10 Subjective workload scores across the six dimensions of the NASA-TLX (M, manual; A, automated).

DESIGN RECOMMENDATIONS FOR FUTURE USER NEEDS

The benefit of adopting a Driver-Initiated systems design is that an element of command and control remains within the driver's grasp (Banks and Stanton, 2014). This means that rather than becoming a passive monitor of the system (e.g. Byrne and Parasuraman, 1996) the driver remains an active supervisor (thus adopting a DM role – see Chapter 2 for a discussion). This means that although the status of the driver within the control-feedback loop has changed, a Driver-Initiated design could prevent against the disintegration of driver–vehicle links within the control-feedback loops (Banks and Stanton, 2014; Banks et al., 2013). On the basis of the results presented by Banks and Stanton (2015c), it is clear that the prototype system of Driver-Initiated Automation used in this study required further development in order to improve ratings of driver trust and workload. Although it was hypothesised that a Driver-Initiated automated control system could protect against the occurrence of out-of-the-loop performance problems often cited within the wider literature (e.g. Billings, 1988; Endsley and Kaber, 1997), it is not clear whether or not a command–control relationship between the driver and the automation will be successful. However, it is proposed that such a system may be successful if appropriate design modifications are made. These are based upon both improving the transparency of systems design through HMI feedback and setting more appropriate system limits that eliminate the likelihood of the automation behaving in an inconsistent manner (e.g. remove the potential for unsafe lane offerings and unexpected lane changes). Banks and Stanton (2015c) proposed the following systems design modifications:

1. Remove the capability of the automated system to offer an automatic overtake. Although the occurrence of unsafe lane offerings was low (Table 9.1), it highlights an important facet in the development and maintenance of driver–automation cooperation (Hoc et al., 2009). Removing the overtake offering will mean that drivers will need to continue to rely upon their

own judgement in initiating complex driving manoeuvres, with the knowledge that a background system of automation is capable of overriding the driver (i.e. it will not change lanes if a fast-moving vehicle occupies the intended lane) if the manoeuvre cannot be completed safely. In its current state, the prototype used in this study did not reach an acceptable standard of combining human decision-making with automated decision-making (Madhaven and Wiegmann, 2007). According to Moiser and Skitka (1996), human–machine decision-making should result in a high-performing control system that enhances the quality of joint performance. However, the conflict that existed between the driver and the automated system highlights that a thorough appraisal of driver decision-making processes relating to the execution of an overtake manoeuvre had not been completed. Madhaven and Wiegmann (2007) argue that it is essential that such an analysis be completed if automated support systems like this are to be a success. Systems developers may be underestimating the power of 'trust' in determining the success of human–automation performance (Lee and See, 2004; Sheridan and Ferrell, 1974; Walker et al., 2015). Removing the offer of an automatic overtake could provide one solution as performing an automatic overtake manoeuvre is already driver-initiated regardless of its presence. In other words, the offering is meaningless given that as soon as drivers express their intent to change lane, as signalled by manipulation of the directional indicators, the automated system takes over. The removal of the lane change suggestion would therefore not reduce the overall level of automation.

2. Remove the capability of the automated system to change more than one lane at a time. This would mean that in order to travel from Lane 1 to Lane 3 of a highway, two driver-initiated lane changes would be needed. Although only 9% of all overtake manoeuvres resulted in an unexpected lane change, it highlights the potential for such system behaviour to occur in the first place. System behaviour that is unwanted is not only likely to affect ratings of driver trust and workload, but also have the potential to affect the safety of other road users. For instance, an automated manoeuvre could indirectly lead to a road traffic accident if any other vehicles on the network take evasive action to avoid the host vehicle.

3. Improve the HMI feedback provided to the driver during manual driving. Presently, the symbology relating to system availability leads to driver confusion over system state. Although the occurrence of this form of mode confusion may reduce as drivers become more familiar with the system (Larsson, 2012), more transparent HMI content could reduce the frequency of occurrence further (Stanton et al., 2011).

4. Improve the warning system used to encourage drivers to regain control of the vehicle. With nearly one quarter of the sample used in this study failing to place their hands back on the wheel in safety critical driving situations, the current system may fail to maintain the safety of both the driver and other road users. While Noujoks et al. (2014) support the use of visual–auditory takeover requests as featured on the prototype used in the current study,

more explicit warning was needed in this instance. It has been suggested that screen-mounted LEDs may capture attention more quickly (Noujoks et al., 2014). Even so, visual–auditory warnings are more favourable than purely visual warnings as people tend to react more quickly to auditory signals despite preferring visual warnings (e.g. Shelton and Kumar, 2010). Noujoks et al. (2014) found that reaction times to system failure (i.e. time between the warning and the driver regaining control of the vehicle) was between 0 and 5 s for visual–auditory in comparison to 0 and 20 s for purely visual warnings. Of course, drivers failing to regain control of the vehicle may have been attributable to a natural curiosity of discovering system capabilities and limits (e.g. risk compensation; Wilde, 1994).

5. Improve the design of system initiation. Analysis of the interview data revealed that many of the issues encountered by drivers were due to inappropriate control location. For example,

'How do I? How does it switch on? Do I push it up? Oh no, that's the indicator...'

'I can't set the cruise control because I don't know where you turn it on...'

For the current prototype, the control stork was located out of the driver's main field of view, meaning that in some instances, drivers were actively trying to look for the control, which was located on a steering wheel stork underneath the traditional indicator mechanism. These initiation problems could be resolved by moving the location of the automated controls, within easy reach and sight of the driver. Many vehicle manufacturers use steering-mounted controls to engage ACC (e.g. BMW, 2014; Jaguar Land Rover, 2014), which would be a more appropriate location for Driver-Initiated Automation control systems.

Notably, these recommendations are applicable only to the automated feature that was tested.

SUMMARY AND CONCLUSIONS

It was found that despite putting drivers in control of the automated systems by enabling them to accept or ignore behavioural suggestions (e.g. overtake), there were still issues associated with increased workload and decreased trust. Trust and workload are important concepts to consider in the future implementation of higher level autonomy because inappropriate levels of trust may lead to disuse (i.e. drivers reject the potential benefits of the system) or misuse (i.e. drivers become complacent; Parasuraman et al., 1993). In addition, a 'negative' first-time experience in using the system could lead drivers to reject the system completely (i.e. not use it even when it becomes available; Sheridan, 1988). This means that in order for drivers to experience the full benefit afforded by the automation of longitudinal and lateral control, they must have appropriate levels of trust in system operation (Lee and See, 2004).

In order for the implementation of the automation of longitudinal and lateral control to be a success, the driver must be comfortable with the degree of control transfer given to the system. Interestingly, an investigation by Larsson (2012) surrounding the usage and perceived issues of vehicles equipped with traditional ACC technology found that despite previous concerns, the very fact that ACC is not a perfect system means that intermittent changes in control between manual and automated driving is actually beneficial to performance because the driver remains actively in-the-control-loop (Bryne and Parasuraman, 1996; Stanton and Marsden, 1996; Stanton and Young, 2005; Walker et al., 2015).

There are still concerns that after prolonged exposure to highly automated driving, driver desensitisation may occur resulting in a lack of task engagement. If this happens, manual override in unanticipated and unexpected events will be difficult to manage, increase workload and stress as well as create surprise or startle effects (Sarter et al., 1997). For example, Merat and Jamson (2009) and Young and Stanton (2007b) have shown that driver response times to unexpected hazards increase by 1.0–1.5 s when driving with ACC in comparison to manual driving. Notably, this increased response time was for well-motivated and alert drivers. This has previously been attributed to cognitive underload (Vollrath et al., 2011; Young and Stanton, 2002), reduced responsibility (Farrell and Lewandowsky, 2000) and the cost of control transfer between automated and manual control (Funke et al., 2007). These are issues that may become more prevalent as drivers become more familiar with the mode of system operation and experience months or years of high reliability. These remain important areas of future research.

To sum, this chapter builds upon Chapters 7 and 8 by providing a thorough appraisal of the role of the driver within a Driver-Initiated system of automation. Where Chapter 8 revealed evidence of driver–vehicle coordination problems, this chapter advances our understanding of system usability further by highlighting a number of design weaknesses that may contribute to the occurrence of driver–vehicle coordination issues. For example, unexpected system behaviour leading to automation surprise may result from system initiation issues in some cases.

10 Distributed Cognition in the Road Transportation Network

A Comparison of 'Current' and 'Future' Networks

INTRODUCTION

Up until now, this book has been concerned with analysing the impact of differing levels of automation on the role of the driver using the theoretical underpinnings of Distributed Cognition. While this is extremely important, it only represents part of the overall road transportation network. In fact, the functioning of the road transportation network is based upon an infinite number of complex interactions and interdependencies between multiple system agents at a number of levels (Salmon et al., 2014). These include system agents within the road environment (RE; e.g. drivers, pedestrians and vehicles), traffic management centres (TMCs; e.g. traffic management operator and road traffic officers) and external agencies (EAs; e.g. emergency services and radio stations). While these categories of system agent are typically analysed independently from one another, a more holistic sociotechnical systems approach would apply the Distributed Cognition approach at the macro-level.

Some of the tools and techniques used within Phase 1 of the Systems Design Framework offer a means to explore and model Distributed Cognition of the entire driving transportation system. To recap, Phase 1 of the Systems Design Framework is concerned with modelling the behaviour of a system. In order to do this, the framework proposes that researchers complete three steps as shown in Figure 10.1. Firstly, a design concept or prototype system needs to be defined so that system boundaries and limits can be recognised. Secondly, allocation of system function enables the 'workload' of a system to be shared among system agents. These system agents can be both human and non-human given the principles of DSA (Stanton et al., 2006, 2015). Thirdly, it is possible to visualise the interaction and communication that occurs between these system agents using OSDs (Brooks, 1960; Kurke, 1961).

Of course, the Systems Design Framework was initially intended to explore Distributed Cognition of micro-level systems. In order to apply it to a much broader system, a macro-cognitive approach is needed. Thus, in order to satisfy Step 1 of the framework, the system in question needs to be defined. This requires along with an

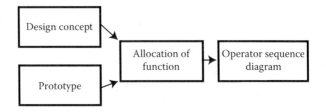

FIGURE 10.1   Phase 1 of the Systems Design Framework.

appraisal of 'who' (both human and non-human) forms this system. The allocation of system function at a macro-level would largely be inappropriate in a driving context. This is because rather than having a single shared goal, the system operates with multiple goals (Stanton, 2014a,b). However, the modelling of system interaction between system agents is possible and can provide an extensive qualitative overview of Distributed Cognition within a wider network. Unlike the Systems Design Framework that utilise OSD representations, to explore Distributed Cognition at a macro-level, the visual representations afforded by techniques such as EAST (Stanton et al., 2013; Walker et al., 2006, 2010) paradigm are more appropriate. EAST utilises task, social and information networks to model Distributed Cognition of complex sociotechnical systems (Stanton, 2014a,b). These representational mediums will therefore be utilised in an effort to portray macro-level Distributed Cognition in the road transportation network.

DISTRIBUTED COGNITION IN THE TRANSPORTATION NETWORK

As explained in the introduction of this chapter, the functioning of the road transportation network is based upon complex interactions and interdependencies between multiple system agents at a number of levels that are not always apparent. Therefore, the first step in modelling Distributed Cognition within the road transportation network is to identify the system agents that constitute it.

IDENTIFICATION OF SYSTEM AGENTS

While it may be impossible to show every possible connection between all of the individual agents in the road transportation network within a visual representation, it is possible to provide an overview of how different classes of agent work together within a shared space (i.e. the road network). For the purposes of this analysis, a total of 21 classes of system agents were identified spanning three broad categories: RE, TMCs and EAs. Table 10.1 introduces the agents that make up these categories as well as provides a description.

TASK NETWORK

A task network that represents the entirety of the transportation network would need to include tasks reflecting the role of all agents within the RE, TMC and EA, as identified in Table 10.1, as well as their interdependencies. Figure 10.2 provides a high-level description of some of the tasks associated with each of these agents, and

TABLE 10.1

Agents Involved within the Road Transportation System

Category	Subcategory	Agent	Description
Road – this category represents all agents that are present within the road environment	Drivers	Host driver Host passenger Other drivers Other passengers	The categories of individuals occupying vehicles
	Vehicles	Host car Other cars Services/goods vehicles Emergency vehicles	The categories of traffic using (or potentially using) the road network
	External roadside equipment	Traffic monitoring equipment Traffic management equipment	For example CCTV cameras and induction loops e.g. traffic lights and Variable Message Signs (VMSs)
		Vulnerable road users	For example cyclists and pedestrians
TMC – this category represents all agents that have direct access to information relating to the overall traffic situation		TMC operator	Responsible for managing traffic
		CCTV applications	Controls the TMC's CCTV cameras
		Urban traffic management control (UTMC) applications	Collects data relating to road environment (e.g. vehicle counts)
		Police CCTV personnel	Monitor CCTV for crime, assisting police operations
EAs – this category represents all agents that both share and receive information relating to traffic situations		Radio stations	Distribute information to traffic and other agents
		Information providers	Provide additional information (e.g. Met Office, Highways England)
		Other transport control centres	Includes other road TMCs as well as public transport control centres (e.g. Bus)
		Emergency services control centres	Manage emergency service operations
		Traffic data distribution services	Dissemination of information to traffic and third parties

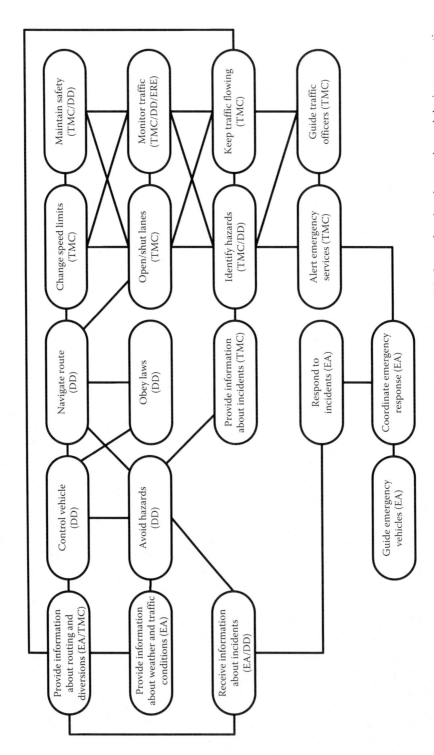

FIGURE 10.2 Task network for the road transportation system showing the agents responsible for performing these tasks, and the interconnections that exist between them.

the interconnections that exist between them. The task network should be viewed as continuous process to reflect the fact that the transportation network is always active. Notably, the distinct role of DD is used instead of 'Driver' as we have already shown in Chapter 2 that multiple driver roles exist. A DD is more representative of the role of the driver within a conventional transportation network as both the physical and the cognitive tasks associated with driving remain the full responsibility of the human operator.

SOCIAL NETWORK

To better understand the structure of communications that occur between the system agents identified in Table 10.1, a social network representation was constructed. Figure 10.3 presents an overall indication of potential communication patterns among agents within a conventional transportation network. In some instances, this communication will be fairly obvious (e.g. gesture between drivers such as flashing lights, radio broadcasts sharing traffic updates). In other instances, communication could be 'invisible' to agents located within the road environment (e.g. UTMC data can be used to change behaviour of traffic management equipment).

Regardless, the social network clearly indicates that relationships exist across and between broad categories of system agents within the road transportation network. Thus, agents in the RE, TMC and EA will at some point share data with one another – even if this occurs unknowingly.

INFORMATION NETWORK

The types of information that may be shared within the road transportation network can be pictorially presented via information networks (Stanton et al., 2013; Walker et al., 2006, 2010). Information networks detail aspects of communication that underpin the foundations of the system. In order to ensure the effective functioning of the road transportation network, a plethora of information must be captured, digested and shared between all of the agents involved. Figure 10.4 combines driver-orientated knowledge (Banks and Stanton, 2017) and TMC-orientated knowledge (Price, 2016) to produce an information network representing the conventional transportation network. The network includes 53 nodes and 60 links. Key informational nodes (i.e. ones most important to effective system functioning) were identified as any node with more than 4 connections. With this in mind, the following nodes were identified as being most important in successful system functioning:

1. *Traffic Type* including vehicles, pedestrians and public transport or services
2. *Traffic* considering the properties of road users, such as speed and route
3. *Infrastructure* considering physical aspects such as road type, capacity and lane markings
4. *Risk Assessment* including hazard type, previous experience and environmental conditions
5. *Strategy* considering the action to be taken in order to manage a scenario
6. *Signage* considerings the meaning and information presented on road signs including speed, warnings and instruction

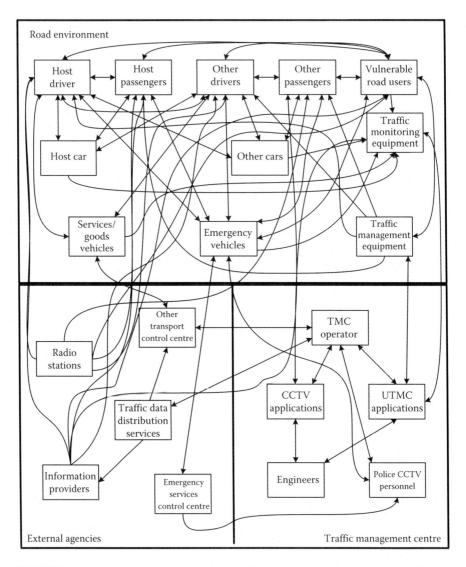

FIGURE 10.3 Social network representation of macro-level communications within the road transportation system.

A COMPARISON OF CONVENTIONAL AND CAV TRANSPORTATION NETWORKS

Distributed Cognition within the road transportation network will become even more prevalent as the network becomes increasingly 'connected' in the advent of CAVs. CAV networks are made up of Vehicle-to-Vehicle (V2V) and Vehicle-to-Infrastructure (V2I) communication streams, which are achieved through the use of Dedicated Short-Range Communication (DSRC) and Global Satellite Positioning

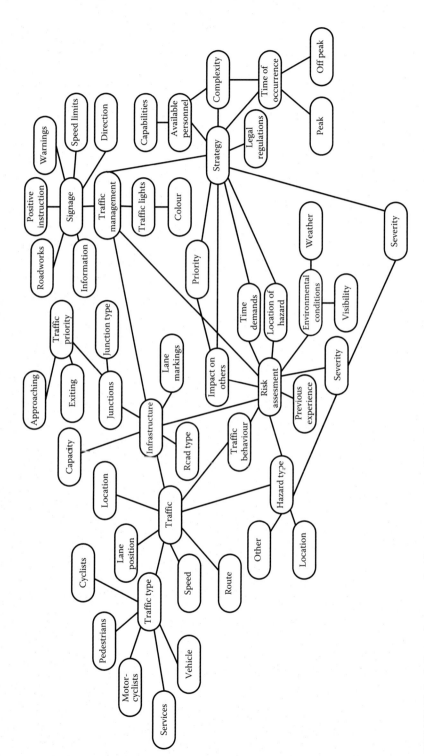

FIGURE 10.4 Information network reflecting macro-level Distributed Cognition of the road transportation system (note that the 'severity' node has been duplicated to disentangle links).

(GPS) sensors. DSRC represents on-board sensor units (in the case of V2V) and external roadside units (in the case of V2I).

The potential impact of CAV functionality upon the transportation task network is shown in Figure 10.5. While the task network does not dramatically change, the agents capable of performing these individual tasks increase. Notably, the authors have included an additional role for the driver (DM). In Chapter 2, we showed that the role of DM reflects the supervisory role that a driver can adopt when the physical and the cognitive tasks associated with driving become automated. The DD remains in the representation to reflect the potential for mixed traffic driving scenarios (e.g. both automated and manual vehicles occupy the road).

In order for the road transportation network to work effectively with the addition of CAV functionality, a greater degree of interaction is required between the social agents. Essentially, CAV points to a future whereby most, if not all, of the potential communication links between vehicles and infrastructure in the RE (see Figure 10.3) may be connected. This is reflected in the CAV social network represented in Figure 10.6. The new links, shown in bold, represent the communication afforded by on-board sensor units (i.e. DSRC) as well as external roadside units (i.e. Variable Message Signs). Thus, intelligent devices falling within the 'Traffic Management Equipment' category (e.g. Variable Message Signs) as well as on-board vehicle sensors (e.g. DSRC and GPS) would be able to send, capture and retransmit data back into the network to both human and non-human agents. One way of thinking about this is to imagine that camera-based technologies and radar enable the vehicle to 'see' (i.e. vision systems that process video data) while DSRC enables the car to 'talk' (i.e. transmit data to other vehicles and infrastructure) and 'listen' (i.e. receive data from other vehicles and infrastructure).

Additional analysis was performed using AGNA™ (version 2.1; Benta, 2005) to compare and contrast network metrics relating to the conventional and CAV social networks. Of most interest was the calculation of sociometric status to assess agent prominence within the networks (Houghton et al., 2006; Salmon et al., 2014). The results, shown in Table 10.2, indicate that 'human' agents play the most prominent role within a conventional network while 'non-human' agents become increasingly prominent within a CAV network alongside their 'human' counterparts. To enable us to see the shift in agent prominence occurring as intelligent functionality is added to the network more clearly, only agents holding different sociometric status values between the social networks are highlighted.

This result implicates the need to further explore the intricate relationships that exist between human and non-human agents within the road transportation system as a whole in order to ensure successful implementation of CAV. While this relationship will continue to evolve as autonomous functionality increases, it is clear that users of the transportation network will always require some level of interaction with non-human agents in order to understand how well the overall system network is operating. Thus, rather than CAV eliminating Human Factors issues from the transportation network, it could actually increase demand as we must consider the types of activities human and non-human agents will engage in, and their interdependencies.

With regards to the knowledge network that may exist for a CAV network, examples of the types of information that can be shared via V2V and V2I are outlined in

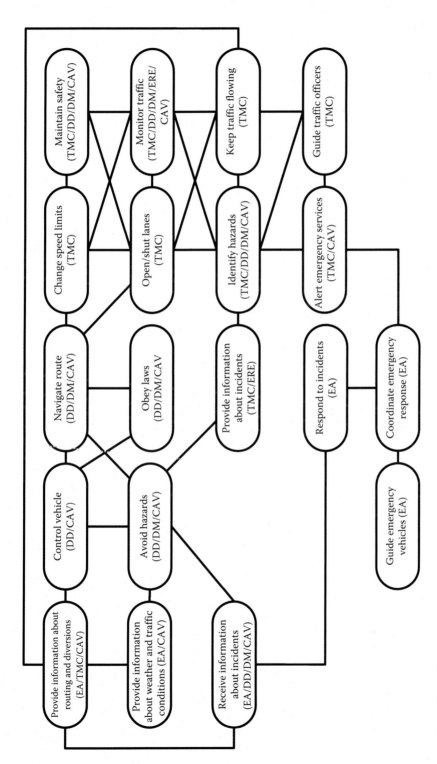

FIGURE 10.5 Task network for future CAV networks in the road transportation domain.

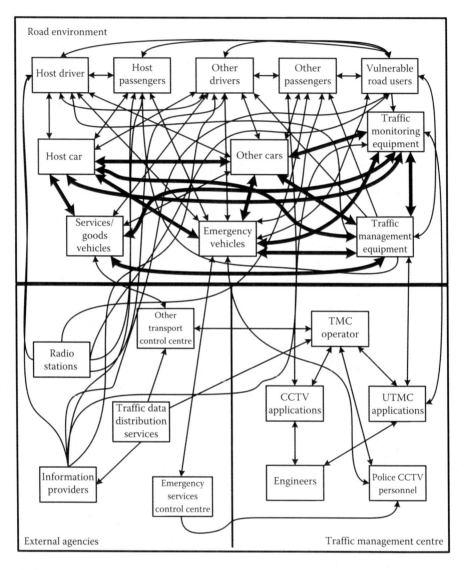

FIGURE 10.6 Social network representation of future CAV networks in the road transportation domain.

Table 10.3 (Society of Automotive Engineers J2735; SAE, 2016). These informational elements have been added to the knowledge network to reflect CAV functionality (see Figure 10.7). This newly extended network has 17 additional nodes and 25 additional connections in comparison to the conventional network representation shown in Figure 10.4. This means in total the network has 71 nodes and 189 edges. In addition to the key nodes identified for the conventional network in Figure 10.4, 'connective functionality' is unsurprisingly the most important informational node within a CAV network.

TABLE 10.2

Contrasting Sociometric Status for Agents Involved in Conventional and CAV Networks

	Conventional Network	CAV Network
Host driver	0.85	0.75
Host passengers	0.55	0.55
Other drivers	0.85	0.85
Other passengers	0.60	0.60
Host vehicle	0.35	0.75
Other vehicles	0.35	0.70
Service/goods vehicles	0.40	0.75
Emergency vehicles	0.50	0.85
Traffic management equipment	0.50	0.70
Traffic monitoring equipment	0.55	0.75
Vulnerable road users	0.65	0.65
TMC operator	0.50	0.50
CCTV applications	0.30	0.30
UTMC applications	0.40	0.40
Police CCTV personnel	0.20	0.20
Engineers	0.20	0.20
Radio stations	0.30	0.30
Information providers	0.25	0.25
Other transport control centres	0.25	0.25
Emergency services control centres	0.10	0.10
Traffic data distribution	0.20	0.20
Mean	*0.42*	*0.50*

TABLE 10.3

Information That Can Be Sent, Captured and Retransmitted Back into the Road Transportation Network via DSRC and GPS

V2V Communication	V2I Communication
Speed	Position
Position	Signal phase and timing
Heading	Local map information
Yaw rate	GPS corrections
Path history	Road condition
Acceleration	Weather
GPS corrections	

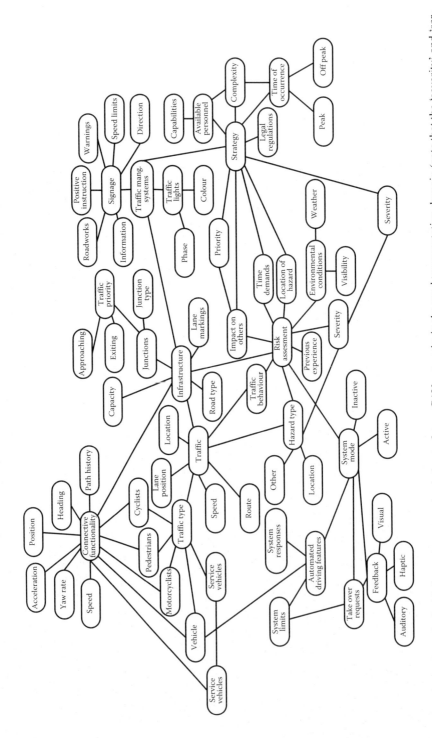

FIGURE 10.7 Hypothetical information network relating to future CAV networks in the road transportation domain (note that the 'severity' and 'service vehicle' nodes have been duplicated to disentangle links).

It is important to note that Figure 10.7 is by no means exhaustive. The very nature of CAV means that the transportation network opens up considerably to services that seek to improve the driving experience. For example, automatic allocated parking could see vehicles seeking out parking spaces within the vicinity and either direct drivers to them, or in the case of autonomous vehicles, drive there independently. This is not captured within the current representation. However, the representations presented in this chapter do provide a basis for discussion into (a) the agents that play a role in our road transportation system, (b) how these agents are connected to one another and (c) the types of communication or information that is shared within the network.

DISCUSSION

Macro-level representations of Distributed Cognition within the road transportation network have shown that with the addition of CAV functionality, an already complex sociotechnical system is set to become even more complex. This becomes problematic if the system itself shows any network resilience issues. This is because failure in one part is likely to have a substantial knock-on effect in others.

One threat posed to CAV networks is cybersecurity and the risk of 'car hacking'. Frost and Sullivan (2014) identified over 50 vulnerable points in which vehicle security could be compromised. Since then, both the media and the academic research have shown numerous instances whereby hackers have overridden vehicular controls and functions within CAV platforms (Checkoway et al., 2011). This has ranged from the hacking of keyless entry systems to more serious breaches of security that have compromised critical safety features (e.g. Checkoway et al., 2011). Thus, network resilience within automotive cybersecurity, understandably, is a rapidly growing field.

Other considerations that could affect the resilience of the CAV network include maintenance of the infrastructure itself. The European Road Transport Research Advisory Council (ERTRAC, 2015) state that traffic signs, signals and road markings must be visible and in good order to support safe and reliable automation functioning. In addition, the Organisation for Economic Cooperation and Development (OECD) recognise that fully autonomous vehicles require precise digital representations and maps of their environment that should be sophisticated enough to deliver information relating to the location of potholes, debris, fog and so on (Lavrinc, 2014) as these factors all affect potential usage patterns and the need for a driver to regain control of the vehicle if and when required.

Implementation routes for CAV are also a hot topic of debate however, one that will impact upon network dynamism greatly. Retrofitting the existing transportation network to accommodate CAV is likely to be extremely expensive. One solution for existing infrastructure is dedicated lanes for CAV. Dedicated lanes for platooning, however, have been found to negatively affect the behaviour of drivers in conventional, non-automated, vehicles. Gouy et al. (2014) found that for manual driving, time headways (THWs) to a lead vehicle was significantly, negatively, affected when drivers were in the vicinity of a platoon keeping short headways (THW = 0.3 s). In many instances, manual drivers maintained a THW of <1 s, which is considered below the safety threshold. These findings add to the growing body of literature that suggests that driving in a platoon can alter the perception of safe driving headways and subsequently lead to behavioural adaptation of manual drivers in keeping short THW (Gouy et al., 2014). Thus, while we remain in an era of mixed traffic driving (i.e. manually driven vehicles and CAV), it is important to consider how these vehicles will interact within a shared space.

CONCLUSIONS

Regardless of CAV implementation routes, issues associated with privacy and security and the many other factors that serve as potential barriers to successful implementation. Vehicle automation and the connected services which accompany it are set to revolutionise the way in which people interact and behave within the road transportation network. The representations afforded by EAST can provide a useful insight into the complex interactions that may take place between system agents. Only when a functioning CAV network exists can Distributed Cognition of the road transportation network (and the EAST models) be verified and validated.

11 Summary of Findings and Research Approach

INTRODUCTION

Vehicle automation is a key international topic of interest for Human Factors. In particular, the impact of automation upon driver behaviour is an extremely active research area worldwide and one that is far from fully resolved. The United Kingdom, Europe, North America, China, Japan and Australasia are all heavily involved in the design, development and in some cases manufacturing of these sorts of technologies. With this in mind, the work presented in this book represents pioneering research into the field of automobile automation – demonstrating how established Human Factors techniques can be applied to this growing domain. It disseminates some of the very latest findings to some of the most pressing real-world questions about the impact of autonomous driving features on the driving system as a whole, which is important because vehicle manufacturers and suppliers are acutely aware of the behavioural challenges that lie ahead.

NOVEL CONTRIBUTIONS OF THIS BOOK

APPLICATION OF DISTRIBUTED COGNITION TO DRIVING

The overall aim of this book was to investigate Distributed Cognition in automated driving systems, with particular emphasis upon how the addition of automation into the driving task affects traditional driver–vehicle interaction patterns. It adds to the growing body of literature that stipulates that knowledge does not lie solely within the individual but rather, knowledge lies within all agents involved within a system network (Hutchins, 1995a; Stanton et al., 2006). Thus, by adopting this systems approach, we have seen how Distributed Cognition can be used to identify the allocation of system function between the driver and automated subsystems in which they interact.

DEVELOPMENT OF A FRAMEWORK TO EXPLORE DISTRIBUTED COGNITION

The authors introduced a novel two-phase Systems Design Framework to explore driver–vehicle interaction using a mix of qualitative and quantitative research methodologies of substantial pedigree. The first phase seeks to model system behaviour to give us a better understanding of Distributed Cognition in driving using traditional task analysis techniques while the second phase seeks to validate these representations using specifically chosen extension methodologies. Empirical methods can also help identify design weaknesses, beyond technological components, that require modification.

Overall, the Systems Design Framework has provided the authors, systems engineers and designers with a foundation to design and conduct research into the Human Factors implications of increasing levels of automation within the driving task. The following sections discuss the specific contributions of some of the tools and techniques used to deliver the findings of this book.

Operator Sequence Diagrams

While in the automotive industry, the use of sequence diagrams has typically been based upon modelling the interaction that occurs between technological components, the Systems Design Framework takes into account all of the agents involved in the driving task (both human and non-human). This simple, yet effective, addition of the driver to otherwise technical representations of system operation has provided non-Human Factors–trained personnel a basic understanding of Distributed Cognition within driving. This is with particular regard to how tasks are shared among system agents (allocation of function), how information within a system is disseminated back to the driver and also as a means to identify gaps, or missing links, between the driver and the automated system.

Collection of Driver Verbalisations

While the analysis of driver verbalisations can be labour intensive (Vitalari, 1985), it has been proven to provide rich insights into how individual drivers behave at varying levels of automation. Thus, while traditional task analysis and modelling can reveal an insight into how system agents may interact or communicate with one another, verbalisations can act as a tool to validate or dispute the accuracy of these descriptions using qualitative analysis techniques. Relying upon quantitative data alone to meet the aims and objectives of this research would have resulted in an incomplete representation of the issues relevant to driving automation. To fully appreciate how automation will impact upon driver–vehicle interaction patterns, the combination of qualitative and quantitative research methodologies is vital.

Qualitative thematic analysis was the primary data analysis technique used in the research presented in this book. Such an approach is highly flexible and can be adapted to meet the needs of the research question. It can provide useful insights that would otherwise be difficult to obtain from purely quantitative data collection. In general, the results are very accessible to many people. On the downside, it is often heavily criticised for not being subject to the same scientific rigour of quantitative analysis due to its subjectivity. However, the novel application of network analysis metrics to otherwise qualitative system models and driver protocols provides an exciting area of future research. This is one way of quantitatively analysing qualitative data.

Even so, Flyvbjerg (2011) argues that qualitative research practice presents no greater bias than those inherent in other research methodologies as researchers take great care to ensure transparency and reliability. With so many questions remaining to be answered relating to what the driver is actually doing within an automated driving system, VPA can provide researchers with a means to gain insight into 'how' and 'why' drivers use automation in the way that they do. Thematic analysis of driver verbalisations in Chapter 5 revealed that while automation in emergencies does not alter traditional driver decision-making pathways (i.e. the processes between hazard detection and response remain similar regardless of the level of automation within the driving task), it does appear to significantly weaken, and in some cases, remove the links between information-processing nodes. This disruption to the decision-making process is an important consideration for the design of future automated technologies as it represents an unexpected outcome of automation implementation.

Similarly, analysis of driver verbalisations in Chapter 8 revealed evidence of driver–vehicle coordination problems during higher levels of automated driving. The type of coordination problems experienced were dependent upon the level of experience that drivers had in using the system. This was an unexpected finding for system developers who had overlooked the potential for miscommunication to occur between the driver and the automated system. Again, driver verbalisations provided insight into the perception of system usability, trust and workload in Chapter 9 enabling the formation of design recommendations to improve HMI design and system transparency.

The main stimulus used to elicit driver verbalisations was exemplar questions from the CDM (Klein et al., 1989). This is traditionally a retrospective interview. However, the associated disadvantages of using the CDM, such as increased vulnerability to memory decay, was highlighted in Chapter 4. To address this issue, recommendations were made by the authors to use CDM in conjunction with a 'freeze probe' technique to address issues of retrieval failure. This alteration to delivery was extremely successful in generating richer insights into driver decision-making in Chapter 5.

Network Analysis

The application of quantitative network metrics to graphical representations within Phase 1 of the Systems Design Framework has proven to be an invaluable analysis tool in highlighting the role of the driver within an automated driving system. Network metrics provide a means to better understand the role and responsibilities of different system agents at varying levels of automation.

In addition, the novel application of network analysis to driver verbalisations has shown to be useful in highlighting the inherent differences in the way that drivers process information at varying levels of automation. This book reports findings that implicate the contextual use of automation but also the appropriateness of systems design in order to properly support the role of the driver. The combination of driver verbalisations and network analysis also provides an area of exciting future research as much can be learned from interrogating verbatim in this way.

Driver Simulator Studies

The advantage of using driving simulation over closed-circuit or on-road driving is that it allows for the evaluation of driver responses to autonomous driving features both without any physical risk (de Winter et al., 2012). In addition to this, it allows for the exploration of any behavioural adaptation as a result of new design concepts or prototypes that drivers may experience in later years (i.e. features that are not yet available within a vehicle). There is a general consensus that driving simulation offers a high degree of controllability as well as reproducibility.

The use of the Southampton University Driving Simulator has enabled the exploration of driver decision-making within emergency situations and how this was affected by the introduction of an AEB system. Following the principles of the Systems Design Framework, the studies suggest that both the level and the type of automated assistance within driving emergencies can affect driver decision-making and subsequent response.

On-Road Trials

This book presents findings of some of the first published on-road investigations of driver–vehicle interaction within highly automated driving systems. This represents a significant step forward in terms of the research and development of automated driving features as such studies are usually confined to closed-circuit or simulator studies (e.g. Stanton et al., 2011). Working alongside an engineering team within a major motor vehicle manufacturer provided both technical expertise and access to vehicles.

Of course within this rapidly growing field, on-road trials will soon become more prevalent as diffusion of highly autonomous vehicles is occurring rapidly among vehicle manufacturers. We are beginning to see distinguishable implementation routes as public testing becomes more commonplace, in particular, for urban and highway environments. Google's fleet of autonomous vehicles is limited to cities or areas that have highly detailed, three-dimensional maps. Delphi and Audi test vehicles have generally been constrained to highway driving (Davies, 2015; Fingas, 2015). Similarly, the test vehicles used within the research presented in this book were also confined to highway driving. This may be because highway automation is potentially 'easier' to achieve as highways are uniformly designed and are better maintained than all other roads on the network (International Transport Forum, 2015). This is an important consideration given that in order to support the implementation of automated vehicles, the road transportation network itself must be maintained appropriately. The ERTRAC (2015) states that traffic signs, signals and road markings must be visible and in good order to support safe and reliable automation. In addition, the OECD recognises that fully autonomous vehicles require precise digital representations and maps of their environment that should be sophisticated enough to deliver information relating to the location of potholes, debris and fog (Lavrinc, 2014). It is after all these types of factors that may affect potential usage patterns and the need for a driver to regain control of the vehicle if and when required.

SUMMARY OF RESEARCH FINDINGS

The work presented in this book was structured around three key objectives identified in Chapter 1. Findings are summarised in relation to these aims and objectives.

OBJECTIVE 1: INCREASE THE AWARENESS OF HUMAN FACTORS IN THE DESIGN OF AUTOMATED AIDS

The development of a Systems Design Framework, introduced in Chapter 3, essentially sought to address this key research objective. Rather than isolating the role of drivers and exploring their behaviour using methodologies from the field of applied cognitive psychology, the authors adopt a systems view keeping drivers firmly embedded within a driving context and exploring how the system as a whole functions as the level and the type of automation begin to change. The Systems Design Framework provided a foundation to all of the subsequent investigations reported in this book.

OBJECTIVE 2: ASSESS THE APPROPRIATENESS OF AUTOMATION DEPLOYMENT AND CONTEXT OF USE

The use of the Southampton University Driving Simulator provided the authors with a unique opportunity to address how different design strategies may impact upon driver behaviour in automated systems. Patten (2013) postulated that a non-warning-based system would be most favourable in preserving the traditional role of the driver but was not necessarily the design strategy of choice among vehicle manufacturers. This indeed appeared to be the case because by 2012, the Euro NCAP had recognised 13 AEB systems, only 3 of which had intentionally rejected the use of warnings. The authors wanted to test the idea that the design of AEB could alter or influence driver behaviour. Findings reported in Chapters 5 and 6 suggest that altering AEB design strategies appears to impact upon decision-making processing *and* driver–vehicle interaction. Remembering that the intention of an AEB system is not to remove the driver from the control-feedback loop but instead serve as a 'backup' in case of the driver failing to intervene effectively, findings presented in Chapters 5 and 6 suggest that different design strategies can affect the way in which drivers process information available to them within the environment and their subsequent response to emergency situations. This means that despite significant improvements to road safety (i.e. reduced frequency of accident involvement with the addition of AEB), there are still important aspects of AEB design that should not be ignored (i.e. the appropriateness of system warnings).

OBJECTIVE 3: PROVIDE DESIGN GUIDANCE ON AUTOMATED FEATURES BASED UPON EXPERIMENTAL EVIDENCE

As the level of automation increases, the risk of the driver becoming out-of-the-loop intensifies. This becomes particularly problematic when a transfer of control is required between the automation and the driver. Ensuring that drivers are in a 'fit state' to achieve this is an enduring challenge. The concept of Driver-Initiated Automation was initially introduced in Chapter 7 as a means to keep the driver in-the-loop. Two on-road studies were conducted using the same prototype system (Chapters 8 and 9). Results presented in Chapter 8 provided a useful, yet worrying, insight into how the level of driver experience in using highly automated driving features may lead to 'automation surprises' in the case of unexpected system behaviour. Recommendations for design improvement and development were proposed in Chapter 9 based upon the results of a thematic analysis of driver verbalisations of which issues relating to the transparency of the HMI were highlighted.

Recommendations were put forward in an effort to both improve system transparency and set further functional limits to avoid 'unexpected system behaviour' that would otherwise be caused by human error (e.g. such as in the case of unexpected lane changes, which were caused directly by the driver inaccurately using the controls).

FUTURE AVENUES OF RESEARCH

The Human Factors issues pertaining to vehicle automation have been speculated about since the 1970s (Sheridan, 1970). Simply removing drivers from the control-feedback loop and eliminating their responsibility over safe vehicle operation do not, however, warrant Human Factors completely redundant. Instead, vehicles operating at increased levels of autonomy with 'self-driving' capabilities open up new avenues of investigation. While much of the research conducted thus far has primarily been concerned with the issues relating to partial automation and complex driver–vehicle interaction patterns, less emphasis has been placed upon the considerations vital to successful market implementation of highly and fully automated driving features (Stanton, 2015). The following sections serve to highlight some of the future Human Factors considerations that may influence the uptake of autonomous driving features. Fagnant and Kockelman (2015) caution that without addressing these 'research gaps', we limit the ability to successfully plan and deliver autonomous vehicles into the transportation system.

DRIVER MONITORING

Although subjective measures of driver workload, SA and trust can offer useful and insightful information relating to driver performance, they are not appropriate techniques to monitor behaviour over time in a naturalistic environment. For this reason, future research should concentrate on developing and enhancing more objective measures of driver monitoring that enable non-invasive, real-time monitoring of drivers and what they are doing. Being able to recognise different driver states at higher levels of driving autonomy will enable researchers to explore the most efficient strategies of transferring control.

Of course, as with all methods of driver monitoring, its capability is dependent upon a number of confounding variables including the complexity and sensitivity of calibration and light quality. However, with the continuing trend to develop systems capable of higher levels of autonomy, there is an increased need for communication and coordination between drivers and automation (Sarter et al., 1997). It is therefore imperative that further research into the area of driver monitoring is conducted in order to ensure that drivers remain aware and are able to access all of the information they require to maintain an appropriate level of SA (Endsley, 1995) at given time.

TRUST AND ACCEPTANCE

Driver trust in technology is a potential barrier to market implementation and should be handled sensitively. Vehicle manufacturers are basically asking drivers to trust the systems that they are designing (Walker et al., 2016) despite that in order to use a new system of automation, especially one that relinquishes the driver of full vehicle control, they must put them in a situation of uncertainty whereby they have incomplete knowledge (Lee and See, 2004). Problems arise when drivers fail to appropriately trust a system which leads to disuse, misuse and abuse (Parasuraman and Riley, 1997). In these situations, Merritt et al. (2013) suggest that serious safety implications can arise.

Trust is a dynamic phenomenon that is based upon attitudes, perceptions and beliefs (Walker et al., 2016). In driving automation, distrust can quickly evolve when the perception of system operation does not conform to the expectations that the driver holds in how the system should operate (Zand, 1972). A recent meta-analysis by Hancock et al. (2011) revealed that 'system performance' had the greatest effect on the development of trust. This may be attributable to the fact that drivers are sensitive observers and in order to learn about system functionality, will carefully observe system behaviour and capabilities (Horswill and Coster, 2002). If drivers feel that they can outperform an automated system, they will generally not use the assistance available to them (Kantowitz et al., 1997). Locus of control therefore becomes an important concept (Montag and Comrey, 1987). Drivers who hold an internal locus of control have high levels of perceived behavioural control and may be more likely find it difficult to relinquish complete control to an automated system in contrast to drivers with an external locus of control (Walker et al., 2016).

Encouraging drivers to trust autonomous driving features is, however, a delicate challenge. Trust is not a well-studied phenomenon despite being inherently important for user acceptance in relation to new vehicle technologies (Walker et al., 2016). One negative experience could significantly affect subsequent use. Perception issues such as these could delay implementation (Fagnant and Kockelman, 2015). Acknowledging and understanding public opinions of autonomous vehicles are therefore important because it will affect the extent to which people will accept these new technologies. It will define the way in which vehicle manufacturers develop and market their vehicles of the future (Kyriakidis et al., 2015).

In relation to acceptance, an individual's decision to use any form of automated system is based upon a number of attitudinal factors. These include trust in system operation (Ghazizadeh et al., 2012), perceived usefulness (e.g. Davis, 1989), social influence (e.g. Venkatesh et al., 2003) and workload (e.g. Young and Stanton, 2002). All of these factors combined can influence an individual's acceptance of new technology. 'A priori' acceptability of a technology infers that the evaluation of a technology can be conducted before having had any interaction with it. This means that it is possible to generate information relating to driver expectations of system functionality; technology acceptability looks specifically at perceived usefulness and ease of use (Davis, 1989).

Indeed, much of the Human Factors work relating to SAE Level 4 autonomy (high automation) and beyond has relied upon surveys, questionnaires and simulator studies (Nowakowski et al., 2015). These have given us important insight into both the types of activities people think they engage with when a vehicle is fully automated and also where they would be most likely to use such systems. Kyriakidis et al. (2015) explored the opinion of 5000 people on automated driving and found that respondents would prefer to use fully automated vehicles on highways, during traffic congestion and for specific tasks such as parking. Only 34% of respondents said that they would use automation in city traffic. These contextual perceptions are important because they give an indication of driver preferences and potential usage patterns. We must not overlook that some drivers enjoy manual driving, and with all of these factors in mind, a Dual-Mode Vehicle (equipped with ADASs that range between manual control and fully automated or driverless functionality; Alessandrini et al., 2015) could be a desirable implementation option. This would give the driver a degree of choice in 'who' performs the driving task and therefore likely to be more acceptable. This would also address speculation that early SAE Level 4 models may be able to offer automated modes only under specific driving conditions such as highway cruising or in low speed conditions (Department for Transport, 2015).

Of course, results from surveys should be viewed with some degree of caution as they rely upon an individual's imagination of what a fully automated car of the future will look like. We ultimately 'cannot foresee what machines can be built to do in the future' (Fitts, 1951, p. 7).

However, expert interviews and predictions relating to future implementation of autonomous vehicles have been found to be fairly accurate (Kyriakidis et al., 2015). For example, an early survey conducted by Underwood (1992) sought to explore which automated features would most likely be deployed in North America. Fifty-five leading experts predicted that ACC would be the most popular technology and would achieve 5% market share by 2004 and 50% by 2015. They postulated that Automated Braking and Lane-Keeping Assist would follow suit. Both of these systems are of course available today. Most interestingly, the experts predicted that fully automated vehicles would ever achieve only 5% market share between 2040 and 2075 and unlikely to ever reach 50% market share. This may not necessarily be due to a rejection of the technology but more a reluctance to pay for it. A recent study revealed that out of 17,400 vehicle owners, 37% of them would be interested in having a fully automated vehicle, but this acceptance level fell to 20% when an estimated market price was provided (Power, 2012).

TRAVEL SICKNESS

Vehicles of the future will not necessarily look like vehicles of today. With the need for driver intervention being kept to a minimum while automated systems are engaged, there is theoretically no need for a steering wheel or traditional brake and accelerator pedals. Indeed, concept vehicles such as the Mercedes Benz Future Truck 2025 (Mercedes, 2015) and the Rinspeed XchangeE (Forbes, 2014) suggests that designs are moving away from traditional vehicle interiors to ones that include steering wheels that can be stowed away or slide to the centre of the car, seats that swivel away from the steering wheel and seating configurations that enable face-to-face conversation with other passengers.

With this in mind, the design of the 'driver cockpit' is likely to be very different to what we are used to now (Casner et al., 2016). Diels and Bos (2016) present hypothetical wireframe designs showing three main scenarios of driving automation: one of which represents the DM role and two that represent the DND role (engagement in secondary task and rearward facing seating arrangement).

Interestingly, two of these scenarios still include a steering wheel. However, the design of the 'driver cockpit', however, is an important consideration in terms of both comfort and suitability to engage in other tasks as well as the effect it may have on the driver experiencing travel sickness. According to Sivak and Schoettle (2015), autonomous driving will increase the frequency and severity of motion sickness as drivers assume a DND role and no longer have to monitor the environment. It is widely known that when control over a task diminishes, individuals are left more susceptible to motion sickness because they are no longer able to accurately predict the future motion trajectory of the vehicle (Diels and Bos, 2016). This is why other vehicle occupants are more likely to experience travel sickness in comparison to the DD (Rolnick and Lubow, 1991). This issue will likely be exacerbated when the driver adopts a DND role and engages in a secondary task that obstructs their view of the outside environment leading to visual–vestibular conflict. This is because a static or dynamic image afforded by secondary task engagement (e.g. reading a book or watching an in-vehicle display) will not correspond to the motion of the vehicle (e.g. Kato and Kitazaki, 2008). It therefore seems that all of the scenarios envisaged for SAE Level 4 automation that enable drivers to engage in tasks of their choice will be important factors in the development of motion sickness. Of course, the onset of symptoms is normally within the range of 10–20 min (O'Hanlon and McCauley, 1974). It would be interesting to find out whether this remains to be the case during autonomous driving as it would significantly affect acceptance and use. Drivers may be unable or unwilling to use SAE Level 4 autonomy and due to the occurrence of motion sickness. This would mean that the benefits of technology will not be capitalised on and, if not taken seriously, affect user acceptance and uptake.

Even so, the development of measures that can mitigate against the effects of motion sickness is important lines of enquiry. Diel and Bos (2016) stress that current design concepts and scenarios for SAE Level 4 driving automation fail to consider basic perceptual mechanisms that can cause occupant discomfort. Of course, obvious measures would be to ensure that vehicle occupants have sufficient visual information available to enable them to anticipate future vehicle trajectories. This may also be achieved via artificial enhancement (i.e. augmented reality).

Feenstra et al. (2011) reported a four-fold reduction in airsickness when a future motion trajectory was presented in an aviation simulator study. Although it remains to be seen whether these results transfer to automated driving, it does present an exciting area of future research. The idea of presenting additional visual information to the driver (including but not limited to future motion trajectory) has been previously used to assess the driver's ability to regain control of an automated vehicle when it was reaching its system limits (Weißgerber et al., 2012). This study supported driver SA and improved their ability to regain control of the vehicle. Augmented reality therefore shows some promise in combating both the risk of sickness and reductions in SA discussed above.

In addition, the size and location of any display within the vehicle should be located as close to the line of sight out of the window as possible (Diel and Bos, 2016). This will mean that vehicle occupants can view the display using their central vision and gain information relating to their direction of travel from their peripheral vision. This will assist vehicle occupants in being able to anticipate the future vehicle trajectory. All of these measures will require extensive validation in practice.

STANDARDISATION

There are many 'unknowns' associated with the implementation of automated driving systems into the road transportation network. While traditional legal regimes assume that the individual sitting in the driver's seat is in control of the vehicle, this is not strictly the case for automated vehicles offering some level of 'self-driving' capability. Autonomous driving therefore presents unique legal challenges that are not effectively addressed by traditional regimes. Standardisation is a crucial area for future research effort. This is because an otherwise fragmented approach will hinder implementation at both national and international levels. The implementation of autonomous vehicles can be successful only if different states and countries work within the same jurisdictions. Thus, more research is needed to produce an ethical and legal framework in which autonomous vehicles will flourish. Even so, we are undoubtedly making positive progress towards the implementation of higher levels of autonomy within the driving domain. However, concerns about liability may hinder innovation during these early years (Garza, 2012).

CLOSING REMARKS

This book takes a new theoretical perspective on the changing role of the driver within automated vehicles. The adoption of a systems theoretic approach provides the necessary foundations and methods to explore the non-linearity experienced in complex sociotechnical systems (Walker et al., 2010). It is hoped that the work presented in this book will promote and encourage systems designers and engineers to consider the role of the driver within an automated system network at the earliest design phase so that Human Factors issues can be addressed quickly and effectively. The Systems Design Framework and the methods it employs provide a foundation to do this within both academic and industrial research.

References

Air Transport Administration. 1989. National plan to enhance aviation safety through human factors improvements. Washington, DC: Author.

Alessandrini, A, A Campagna, P Delle Site, F Filippi, and L Persia. 2015. Automated vehicles and the rethinking of mobility and cities. *Transportation Research Procedia* 5: 145–160.

Amditis, A, K Pagle, S Joshi, and E Bekiaris. 2010. Driver-vehicle-environment monitoring for on-board driver support systems: Lessons learned from design and implementation. *Applied Ergonomics* 41 (2): 225–235.

Andre, A and A Degani. 1997. Do you know what mode you're in? An analysis of mode error in everyday things. In *Human-Automation Interaction: Research and Practice*, by M Mouloua and J M Koonce (eds.), pp. 19–28. Mahwah, NJ: Lawrence Erlbaum.

Annett, J. 2004. Hierarchical task analysis. In *The Handbook of Task Analysis for Human-Computer Interaction*, by D Diaper and N A Stanton (eds.), pp. 67–82. Mahwah, NJ: Lawrence Erlbaum.

Artman, H and C Garbis. 1998. Situation awareness as distributed cognition. *Proceedings of 9th Conference of Cognitive Ergonomics*, pp. 151–156. Limerick, Ireland: ECCE.

Ashleigh, M J and N A Stanton. 2001. Trust: Key elements in human supervisory control domains. *Cognition, Technology & Work* 3: 92–100.

Baber, C, N A Stanton, J Atkinson, R McMaster, and R J Houghton. 2013. Using social network analysis and agent-based modelling to explore information flow using common operational pictures for maritime search and rescue operations. *Ergonomics* 56 (6): 889–905.

Bach, K M, M G Jæger, M B Skov, and N G Thomassen. 2009. Interacting with in-vehicle systems: Understanding, measuring, and evaluating attention. *Proceedings of the 23rd British HCI Group Annual Conference on People and Computers: Celebrating People and Technology*, Cambridge, UK, 1–5 September 2009, pp. 453–462. Cambridge, UK: British Computer Society.

Bainbridge, L. 1983. Ironies of automation. *Automatica* 19 (6): 775–779.

Bainbridge, L. 1999. Verbal reports as evidence of the process operator's knowledge. *International Journal of Human-Computer Studies* 51: 213–238.

Bandura, A. 1986. *Social Foundations of Thought and Action: A Social Cognitive Theory*. Englewood Cliffs, NJ: Prentice-Hall.

Banks, V A and N A Stanton. 2014. Hands and feet free driving: Ready or not? *Proceedings of the 5th International Conference on Applied Human Factors and Ergonomics (AHFE)*. Kraków, Poland, 19–23 July 2014.

Banks, V A and N A Stanton. 2015a. Contrasting models of driver behaviour in emergencies using retrospective verbalisations and network analysis. *Ergonomics* 58 (8): 1337–1346.

Banks, V A and N A Stanton. 2015b. Discovering driver-vehicle coordination problems in future automated control systems: Evidence from verbal commentaries. *Proceedings of the 6th International Conference on Applied Human Factors and Ergonomics*. Las Vegas, NV, 26–30 July 2015.

Banks, V A and N A Stanton. 2015c. Keep the driver in control: Automating automobiles of the future. *Applied Ergonomics* 53 (B): 389–395.

Banks, V A and N A Stanton. in press. Analysis of driver roles: Modelling the changing role of the driver in automated driving systems using EAST. *Theoretical Issues in Ergonomics Science*.

Banks, V A, N A Stanton, and C Harvey. 2013. What the crash dummies don't tell you: The interaction between driver and automation in emergency situations. *Proceedings of the IEEE Intelligent Transportation Systems for All Transport Modes 2013*. The Hague, The Netherlands, 6–9 October 2013.

Banks, V A, N A Stanton, and C Harvey. 2014a. Sub-systems on the road to vehicle automation: Hands and feet free but not 'mind' free driving. *Safety Science* 62: 505–514.

Banks, V A, N A Stanton, and C Harvey. 2014b. What the drivers do and do not tell you: Using verbal protocol analysis to investigate driver behaviour in emergency situations. *Ergonomics* 57 (3): 332–342.

Baumeister, R F, K D Vohs, and D C Funder. 2007. Psychology as the science of self-reports and finger movements: Whatever happened to actual behaviour. *Perspectives on Psychological Science* 2 (4): 396–403.

Bella, F. 2008. Driving simulator for speed research on two-lane rural roads. *Accident Analysis and Prevention* 40: 1078–1087.

Benta, M. 2005. Studying communication networks with AGNA 2.1. *Cognition Brain Behaviour* 9: 567–574.

Billings, C E. 1988. Toward human centred automation. In *Flight Deck Automation: Promises and Realities*, by S D Norman and H W Orlady (eds.), pp. 167–190. Moffet Field, CA: NASA-Ames Research Center.

Blana, E. 1996. A survey of driving research simulators around the world. Working Paper 481. Institute of Transport Studies, University of Leeds, http://www.bmw.com/com/en/insights/technology/technology_guide/articles/cruise_control.html.

BMW. 2014. Cruise control. http://www.bmw.com/com/en/insights/technology/technology_guide/articles/cruise_control.html.

Boren, M T and J Ramey. 2000. Thinking aloud: Reconciling theory and practice. *IEEE Transactions on Professional Communication* 43: 261–278.

Braun, V and V Clarke. 2006. Using thematic analysis in psychology. *Qualitative Research in Psychology* 3 (2): 77–101.

Brookhuis, K A and D de Waard. 2006. The consequences of automation for driver behaviour and acceptance. *Proceedings of the 16th World Congress of the International Ergonomics Association*. Maastricht, The Netherlands, 10–14 July 2006.

Brookhuis, K A and D de Waard. 2010. Monitoring drivers' mental workload in driving simulators using physiological measures. *Accident Analysis and Prevention* 42 (3): 898–903.

Brookhuis, K A, D de Waard, and S H Fairclough. 2003. Criteria for driver impairment. *Ergonomics* 46: 433–445.

Brooks, F A. 1960. Operational sequence diagrams. *IRE Transactions on Human Factors in Electronics* HFE-1 (1): 33–34.

Builder, C H, S C Bankes, and R Nordin. 1999. *Command Concepts: A Theory Derived from the Practice of Command and Control*. Santa Monica, CA: Rand.

Bureau d'Enquêtes et d'Analyses. 2012. Final report on the accident on 1st June 2009 to the Airbus A330-203 registered F-GZCP operated by Air France flight AF 447 Rio de Janeiro – Paris. Accessed November 2014. http://www.bea.aero/docspa/2009/f-cp090601.en/pdf/f-cp090601.en.pdf.

Byrne, E A and R Parasuraman. 1996. Psychophysiology and adaptive automation. *Biological Psychology* 42 (3): 249–268.

Camps, J. 2003. Concurrent and retrospective verbal reports as tools to better understand the role of attention in second language tasks. *International Journal of Applied Linguistics* 13: 201–221.

Cantin, V, M Lavalliere, M Simoneau, and N Teasdale. 2009. Mental workload when driving in a simulator: Effects of age and driving complexity. *Accident Analysis and Prevention* 41 (4): 763–771.

Carsten, O, F C H Lai, Y Barnard, A H Jamson, and N Merat. 2012. Control task substitution in semi-automated driving: Does it matter what aspects are automated? *Human Factors* 54 (5): 747–761.

Casner, S M, E L Hutchins, and D Norman. 2016. The challenges of partially automated driving. *Communications of the ACM* 59 (5): 70–77.

Chapanis, A. 1995. Ergonomics in product development: A personal view. *Ergonomics* 38 (8): 1625–1638.

Checkoway, S, D McCoy, B Kantor, D Anderson, H Shacham, S Savage, K Koscher, A Czeskis, F Roesner, and T Kohno. 2011. Comprehensive experimental analyses of automotive attack surfaces. *Proceedings of 20th USENIX Security Symposium*, 8–12 August 2011. San Francisco, CA.

Civil Aviation Authority. 1998. *Global Fatal Accident Review 1980–96 (CAP 681)*. London, UK: Civil Aviation Authority.

Continental. 2014. In the future, who will be doing the driving, the driver or the car itself? Accessed 14 April 2016. http://www.continental-corporation.com/www/pressportal_com_en/general/automated-driving/automated-driving-intro-en.html.

Cuevas, H M, S M Fiore, B S Caldwell, and L Strater. 2007. Augmenting team cognition in human-automation teams performing in complex operational environments. *Aviation, Space, and Environmental Medicine* 78: B63–B70.

Davis, F D. 1989. Perceived usefulness, perceived ease of use, and user acceptance of information technology. *MIS Quarterly* 13 (3): 319–340.

Davies, A. 2015. This is big: A robo-car just drove across the country. Accessed 3 April 2017. http://www.wired.com/2015/04/delphi-autonomous-car-cross-country/.

Davis, F D, R P Bagozzi, and P R Warshaw. 1989. User acceptance of computer technology: A comparison of two theoretical models. *Management Science* 35: 982–1003.

de Waard, D, M van der Hulst, M Hoedemaeker, and K Brookhuis. 1999. Driver behaviour in an emergency situation in the automated highway system. *Transportation Human Factors* 1: 67–82.

de Winter, J C F, R Happee, M H Martens, and N A Stanton. 2014. Effects of adaptive cruise control and highly automated driving on workload and situation awareness: A review of the empirical evidence. *Transportation Research Part F: Traffic Psychology and Behaviour* 27: 196–217.

de Winter, J C F, N A Stanton, J S Price, and H Mistry. 2016. The effects of driving with different levels of unreliable automation on self-reported workload and secondary task performance. *International Journal of Vehicle Design* 70 (4): 297–324.

de Winter, J C F, P van Leeuwen, and R Happee. 2012. Advantages and disadvantages of driving simulators: A discussion. *Proceedings of Measuring Behavior 2012*. The Netherlands: Utrecht, pp. 47–50.

Dehais, F, M Causse, F Vachon, and S Tremblay. 2012. Cognitive conflict in human-automation interactions: A psychological study. *Applied Ergonomics* 43 (3): 588–595.

Department for Transport. 2013. National Travel Survey 2013 Statistical Release. Accessed 5 October 2015. https://www.gov.uk/government/uploads/system/uploads/attachment_data/file/342160/nts2013-01.pdf.

Department for Transport. 2015. The pathway to driverless cars: Summary report and action plan. Accessed 19 September 2016. https://www.gov.uk/government/uploads/system/uploads/attachment_data/file/401562/pathway-driverless-cars-summary.pdf.

Desmond, P A, P A Hancock, and J L Monette. 1998. Fatigue and automation-induced impairments in simulated driving performance. *Transportation Research Record* 1628: 8–14.

Desmond, P A and G Matthews. 2009. Individual differences in stress and fatigue in two field studies of driving. *Transportation Research Part F* 12: 265–276.

Diels, C and J E Bos. 2016. Self-driving carsickness. *Applied Ergonomics* 53: 374–382.

Dingus, T A, A W Gellatly, and S J Reinach. 1997. Human computer interaction applications for intelligent transportation systems. In *Handbook of Human-Computer Interaction*, by M Helander, T K Landauerand and P Prabhu (eds.), pp. 1259–1282. Amsterdam, The Netherlands: Elsevier.

Donmez, B, L N Boyle, and J D Lee. 2007. Safety implications of providing real-time feedback to distracted drivers. *Accident Analysis and Prevention* 39 (3): 581–590.

Dozza, M. 2012. What factors influence drivers' response time for evasive maneuvers in real traffic? *Accident Analysis and Prevention* 58: 299–308.

Driskell, J E and B Mullen. 2005. Social network analysis. In *Handbook of Human Factors and Ergonomics Methods*, by N A Stanton, A Hedge, K Brookhuis and E Salas (eds.), pp. 58.1–58.6. London, UK: CRC Press.

Dzindolet, M T, L G Pierce, H P Beck, and L A Dawe. 2002. The perceived utility of human and automated aids in a visual detection task. *Human Factors* 44: 79–94.

Endsley, M R. 1988. Design and evaluation for situation awareness enhancement. *Proceedings of the Human Factors Society 32nd Annual Meeting*, pp. 97–101. Santa Monica, CA: Human Factors Society.

Endsley, M R. 1995. Toward a theory of situation awareness in dynamic systems. *Human Factors* 37 (1): 32–64.

Endsley, M R. 2006. Situation awareness. In *Handbook of Human Factors and Ergonomics* (3rd ed.), by G Salvendy (ed.), pp. 528–542. Hoboken, NJ: John Wiley & Sons.

Endsley, M R and D B Kaber. 1997. Out-of-the-loop performance problems and the use of intermediate levels of automation for improved control system functioning and safety. *Process Safety Progress* 16 (3): 126–131.

Endsley, M R and D B Kaber. 1999. Level of automation effects on performance, situation awareness and workload in a dynamic control task. *Ergonomics* 42 (3): 462–492.

Endsley, M R and E O Kiris. 1995. The out-of-the-loop performance problem and level of control in automation. *Human Factors* 37 (2): 381–394.

Ericsson, K A. 2002. Towards a procedure for eliciting verbal expression of non-verbal experience without reactivity: Interpreting the verbal overshadowing effect within the theoretical framework for protocol analysis. *Applied Cognitive Psychology* 16 (8): 981–987.

Ericsson, K A and H A Simon. 1993. *Protocol Analysis: Verbal Reports as Data*. Cambridge, MA: MIT Press.

Erlandsson, M and A Jansson. 2007. Collegial verbalisation: A case study on a new method on information acquisition. *Behaviour & Information Technology* 26 (6): 535–543.

European Commission. 2011. White paper: Roadmap to a single European transport area – Towards a competitive and resource efficient transport system. COM 144.

European Parliament and the Council for the European Union. 2009. Regulation (EC) No 661/2009. Concerning type-approval requirements for the general safety of motor vehicles, their trailers and systems, components and separate technical units intended therefor. Accessed 20 January 2009. http://europa.eu/legislation_summaries/internal_market/single_market_for_goods/motor_vehicles/technical_implications_road_safety/mi0053_en.htm.

European Road Transport Research Advisory Council (ERTRAC). 2015. ERTRAC Automated Driving roadmap, July 2015. Accessed 3 April 2017. http://www.ertrac.org/uploads/documentsearch/id38/ERTRAC_Automated-Driving-2015.pdf.

Fagnant, D J and K Kockelman. 2015. Preparing a nation for autonomous vehicles: Opportunities, barriers and policy recommendations for capitalising on self-driven vehicles. *Transportation Research Part A* 77: 167–181.

Fancher, P, R Ervin, J Sayer, M Hagan, S Bogard, Z Bareket, M Mefford, and J Haugen. 1998. *Intelligent Cruise Control Field Operation Test: Final Report*. Report No. DOT HS 808 849, National Highway Traffic and Safety Administration.

Farrell, S and S Lewandowsky. 2000. A connectionist model of complacency and adaptive recovery under automation. *Journal of Experimental Psychology: Learning, Memory and Cognition* 26 (2): 395–410.

Federal Aviation Administration. 1990. *The National Plan for Aviation Human Factors.* Washington, DC: Federal Aviation Administration.

Federal Aviation Administration. 2003. Standard operating procedures for flight deck crew members. AC No. 120-71A. Accessed 3 April 2017. https://www.faa.gov/documentLibrary/media/Advisory_Circular/AC120-71A.pdf.

Feenstra, P J, J E Bos, and R N H W van Gent. 2011. A visual display enhancing comfort by counteracting airsickness. *Displays* 32: 194–200.

Fingas, J. 2015. Audi's self-driving car is traveling 550 miles to Las Vegas. Published 5 January 2015. http://www.engadget.com/2015/01/05/audi-self-driving-car-drivesto-ces/.

Fitts, P M. 1951. *Human Engineering for an Effective Air Navigation and Traffic Control System.* Washington, DC: National Research Council.

Fitzhugh, E E, R R Hoffman, and J E Miller. 2011. Active trust management. In *Trust in Military Teams*, by N A Stanton (ed.). Aldershot, UK: Ashgate, pp. 197–217.

Fleming, B. 2012. New automotive electronics technologies. *IEEE Vehicular Technology Magazine* 7 (4): 4–12.

Flemisch, F O, C A Adams, S R Conway, K H Goodrich, M T Palmer, and P C Schutte. 2003. *The H-Metaphor as a Guideline for Vehicle Automation and Interaction.* Hampton, VA: NASA Langley Research Center.

Flemisch, F O, K Bengler, H Bubb, H Winner, and R Bruder. 2014. Towards cooperative guidance and control of highly automated vehicles: H-mode and conduct-by-wire. *Ergonomics* 57 (3): 343–360.

Flemisch, F, M Heesen, T Hesse, J Kelsch, A Schieben, and J Beller. 2012. Towards a dynamic balance between humans and automation: Authority, ability, responsibility and control in shared and cooperative situations. *Cognition, Technology & Work* 14: 3–18.

Flyvbjerg, B. 2011. Case study. In *The Sage Handbook of Qualitative Research*, by N K Denzin and Y S Lincoln (eds.), pp. 301–316. Thousand Oaks, CA: Sage.

Forbes. 2014. Self-driving cars can be relaxing, as the Rinspeed XchangE concept shows. Accessed 10 May 2016. http://www.forbes.com/sites/matthewdepaula/2014/02/28/rinspeed-xchange-concept-has-interior-designed-for-autonomous-driving/#25f99b0f7a3b.

Frost & Sullivan. 2014. Cybersecurity in the automotive industry. Published 30 October 2014. Accessed 09 January 2017. http://www.frost.com.

Fuller, R. 2005. Towards a general theory of driver behaviour. *Accident Analysis and Prevention* 37: 461–472.

Funke, G, G Matthews, J S Warm, and A K Emo. 2007. Vehicle automation: A remedy for driver stress? *Ergonomics* 50 (8): 1302–1323.

Gandhi, T and M M Trivedi. 2007. Pedestrian protection systems: Issues, survey, and challenges. *IEEE Transactions on Intelligent Transportation Systems* 8 (3): 413–430.

Garza, A P. 2012. 'Look Ma, No Hands!': Wrinkles and wrecks in the age of autonomous vehicles. *New England Law Review* 46 (581): 581–616.

Gasser, T. 2014. Vehicle automation: Definitions, legal aspects, research needs. *UNECE Workshop: German Federal Highway Research Institute: Towards a New Transportation Culture: Technology Innovations for Safe, Efficient and Sustainable Mobility.* Brussels, Belgium, November 17–18, 2014.

Gasser, T M and D Westhoff. 2012. BASt-study: Definitions of automation and legal issues in Germany. *Road Vehicle Automation Workshop.* Irvine, CA, 24–26 July 2012.

Georgescu, M P. 2006. Driverless CBTC-specific requirements for CBTC systems to overcome operation challenges. In *Computers in Railways X: Computer System Design and Operation in the Railway and Other Transit Systems*, by J Allan, C A Brebbia, A F Rumsey, G Sciutto and S Sone (eds.), pp. 401–413. Southampton, UK: WIT Press.

Ghazizadeh, M, Y Peng, J D Lee, and L Boyle. 2012. Augmenting the Technology Acceptance Model with Trust: Commercial drivers' attitudes towards monitoring and feedback. *Proceedings of the Human Factors and Ergonomics Society 2012 Annual Meeting*, Boston, MA, 22–26 October 2012, pp. 2286–2290.

Godley, S T, T J Triggs, and B N Fildes. 2002. Driving simulation validation for speed research. *Accident Analysis and Prevention* 34: 589–600.

Gorman, J C, N J Cooke, and J L Winner. 2006. Measuring team situation awareness in decentralised command and control environments. *Ergonomics* 49 (12–13): 1312–1325.

Gouy, M, K Weidmann, A Stevens, G Brunett, and N Reed. 2014. Driving next to automated vehicle platoons: How do short time headways influence nonplatoon drivers' longitudinal control? *Transportation Research Part F* 27: 264–273.

Greenberg, J A and T J Park. 1994. The Ford driving simulator. SAE Technical Paper Series No. 940176.

Griffin, T G C, M S Young, and N A Stanton. 2010. Investigating accident causation through information network modelling. *Ergonomics* 53 (2): 198–210.

Grote, G, S Weik, T Wafler, and M Zolch. 1995. Criteria for the complementary allocation of functions in automated work systems and their use in simultaneous engineering projects. *International Journal of Industrial Ergonomics* 16: 326–382.

Grover, C, I Knight, F Okoro, I Simmons, G Couper, P Massie, and B Smith. 2008. Automated emergency braking systems: Technical requirements, costs and benefits – Final 2008. TRL Limited. Accessed 31 January 2013. http://ec.europa.eu/enterprise/sectors/automotive/files/projects/report_aebs_en.pdf.

Gugerty, L J. 1997. Situation awareness during driving: Explicit and implicit knowledge in dynamic spatial memory. *Journal of Applied Experimental Psychology* 3 (1): 42–66.

Habibovic, A, E Tivesten, N Uchida, J Bargman, and M L Aust. 2013. Driver behaviour in car-to-pedestrian incidents: An application of the driving reliability and error analysis method (DREAM). *Accident Analysis and Prevention* 50: 554–565.

Halverson, C A. 1995. Inside the cognitive workplace: New technology and air traffic control. PhD thesis, University of California, San Diego, CA.

Hancock, P A. 2003. Individuation: Not merely human-centred but person-specific design. *Proceedings of the Human Factors and Ergonomics Society* 47: 1085–1086.

Hancock, P A and R Parasuraman. 1992. Human factors and safety in the design of intelligent vehicle highway systems. *Journal of Safety Research* 23 (4): 181–198.

Hancock, P A and S F Scallen. 1996. The future of function allocation. *Ergonomics in Design: The Quarterly of Human Factors Applications* 4: 24–29.

Hancock, P A, D R Billings, K E Schaefer, J Y C Chen, E J de Visser, and R Parasuraman. 2011. A meta-analysis of factors affecting trust in human-robot interaction. *Human Factors* 53 (5): 517–527.

Harris, C J and I White. 1987. *Advances in Command, Control and Communication Systems*. London, UK: Peregrinus.

Hart, S G and L E Staveland. 1988. Development of NASA-TLX (Task Load Index): Results of empirical and theoretical research. *Advances in Psychology* 52: 139–183.

Hedlund, J. 2000. Risky business: Safety regulations, risk compensation, and individual behaviour. *Injury Prevention* 6: 82–90.

Heikoop, D D, J C F de Winter, B van Arem, and N A Stanton. 2016. Psychological constructs in driving automation: A consensus model and critical comment on construct proliferation. *Theoretical Issues in Ergonomics Science* 17 (3): 284–303.

Helton, W S, T H Shaw, J S Warm, G Matthews, W N Dember, and P A Hancock. 2004. Workload transitions: Effects on vigilance performance and stress. In *Human Performance, Situation Awareness and Automation: Current Research and Trends*, by D A Vincenzi, M Mouloua and P A Hancock (eds.), pp. 258–262. Mahwah, NJ: Erlbaum.

Hoc, J M. 2000. From human–machine interaction to human–machine cooperation. *Ergonomics* 43 (7): 833–843.

Hoc, J M, M S Young, and J M Blosseville. 2009. Cooperation between drivers and automation: Implications for safety. *Theoretical Issues in Ergonomics Science* 10 (2): 135–160.

Hockey, G R J. 1997. Compensatory control in the regulation of human performance under stress and high workload: A cognitive energetical framework. *Biological Psychology* 45: 73–93.

Hoedemaeker, M and K A Brookhuis. 1998. Behavioural adaptation to driving with an adaptive cruise control (ACC). *Transportation Research Part F: Traffic Psychology and Behaviour* 1 (2): 95–106.

Hoeger, R, A Amditis, M Kunert, A Hoess, F Flemish, H P Krueger, and K Pagle. 2008. Highly automated vehicles for intelligent transport: Have-it approach. *15th World Congress on Intelligent Transport Systems and ITS America's 2008 Annual Meeting*. New York, NY, 16–20 November 2008.

Hoffman, K A, L M Aitken, and C Duffield. 2009. A comparison of novice and expert nurses' cue collection during clinical decision-making: Verbal protocol analysis. *International Journal of Nursing Studies* 46 (10): 1335–1344.

Hollan, J, E Hutchins, and D Kirsh. 2000. Distributed cognition: Toward a new foundation for human-computer interaction research. *ACM Transactions on Computer Human Interaction* 7: 174–196.

Hollnagel, E. 2001. Extended cognition and the future of ergonomics. *Theoretical Issues in Ergonomics Science* 2 (3): 309–315.

Hollnagel, E, A Nabo, and I V Lau. 2004. A systemic model for driver in-control. *Proceedings of the 2nd International Driving Symposium on Human Factors in Driver Assessment*. Park City, UT, 21–24 July 2004, pp. 86–91.

Hollnagel, E and D Woods. 1983. Cognitive systems engineering: New wine in new bottles. *International Journal of Human Computer Studies* 18: 583–600.

Horswill, M S and M E Coster. 2002. The effect of vehicle characteristics on drivers' risk taking behaviour. *Ergonomics* 4 (2): 85–104.

Houghton, R J, C Baber, R McMaster, N A Stanton, P M Salmon, R Stewart, and G H Walker. 2006. Command and control in emergency services operations: A social network analysis. *Ergonomics* 49: 1204–1225.

Hughes, J and S Parkes. 2003. Trends in the use of verbal protocol analysis in software engineering research. *Behaviour & Information Technology* 22 (2): 127–140.

Hutchins, E. 1995a. How a cockpit remembers its speed. *Cognitive Science* 19 (3): 265–288.

Hutchins, E. 1995b. *Cognition in the Wild*. Cambridge, MA: MIT Press.

Hutchins, E and T Klausen. 1996. Distributed cognition in an airline cockpit. In *Communication and Cognition at Work*, by D Middleton and Y Engestrom (eds.). Cambridge, UK: Cambridge University Press, pp. 15–34.

International Transport Forum. 2015. Automated and autonomous driving regulation under uncertainty. Corporate Partnership Board. https://cyberlaw.stanford.edu/files/publication/files/15CPB_AutonomousDriving.pdf.

Jack, A L and A Roepstorff. 2002. Introspection and cognitive brain mapping: From stimulus-response to script-report. *Trends in Cognitive Sciences* 6 (8): 333–339.

Jaguar Land Rover. 2014. 2015 Land Rover Discovery. http://newsroom.jaguarlandrover.com/en-in/land-rover/press-kits/2014/06/lr_discovery_15my_press_kit/.

Jameson, A. 2003. Adaptive interfaces and agents. In *Human-Computer Interface Handbook*, by J A Jacko and A Sears (eds.), pp. 305–330. Mahwah, NJ: Erlbaum.

Jamson, H, N Merat, O M J Carsten, and F C H Lai. 2013. Behavioural changes in drivers experiencing highly-automated vehicle control in varying traffic conditions. *Transportation Research Part C* 30: 116–125.

Jenkins, D P, N A Stanton, P M Salmon and G H Walker. 2009. *Cognitive Work Analysis: Coping with Complexity*. Aldershot, UK: Ashgate.

Jenkins, D P, P M Salmon, N A Stanton, G H Walker, and L Rafferty. 2011. What could they have been thinking? How sociotechnical system design influences cognition: A case study of the Stockwell shooting. *Ergonomics* 54 (2): 103–119.

Jian, J, A M Bisantz, and C G Drury. 2000. Foundations for an empirically determined scale of trust in automated systems. *International Journal of Cognitive Ergonomics* 4 (1): 53–71.

Kaber, D B and M R Endsley. 2004. The effects of level of automation and adaptive automation on human performance, situation awareness and workload in a dynamic control task. *Theoretical Issues in Ergonomics Science* 5 (2): 113–153.

Kaber, D B, Y Liang, Y Zhang, M L Rogers, and S Gangakhedkar. 2012. Driver performance effects of simultaneous visual and cognitive distraction and adaptation behaviour. *Transportation Research Part F: Traffic Psychology and Behavior*, 15 (5): 491–501.

Kakimoto, T, Y Kamei, M Ohira, and K Muatsumoto. 2006. Social network analysis on communications for knowledge collaboration in OSS communities. *Proceedings of the International Workshop on Supporting Knowledge Collaboration in Software Development*, Tokyo, Japan, 19 September, 2006, pp. 35–41.

Kantowitz, B H, R J Hanowski, and S C Kantowitz. 1997. Driver acceptance of unreliable traffic information in familiar and unfamiliar settings. *Human Factors* 39 (2): 164–176.

Kato, K and S Kitazaki. 2008. Improvement of ease of viewing images on an in-vehicle display and reduction of carsickness. SAE Technical Paper Series No. 2008-01-0565. Warrendale, PA: Society of Automotive Engineers.

Keller, C G, T Dang, H Fritz, A Joos, C Rabe, and D M Gavrila. 2011. Active pedestrian safety by automatic braking and evasive steering. *IEEE Transactions on Intelligent Transportation Systems* 12 (4): 1292–1304.

Khan, A M, A Bacchus, and S Erwin. 2012. Policy challenges of increasing automation in driving. *International Association of Traffic and Safety Sciences Research* 35 (2): 79–89.

Kienle, M, D Dambock, J Kelsch, F Flemisch, and K Bengler. 2009. Towards an H-mode for highly automated vehicles: Driving with side sticks. *Proceedings of the 1st International Conference on Automotive User Interfaces and Interactive Vehicular Applications*, pp. 19–23.

Kirwan, B and L K Ainsworth. 1992. *A Guide to Task Analysis*. London, UK: Taylor & Francis.

Klein, G A, R Calderwood, and D Macgregor. 1989. Critical decision method for eliciting knowledge. *IEEE Transactions on Systems, Man and Cybernetics* 19 (3): 462–472.

Klein, G, D Woods, J Bradshaw, R Hoffman, and P Feltovich. 2004. Ten challenges for making automation a team player in joint human-agent activity. *IEEE Intelligent Systems* 19 (6): 91–95.

Kurke, M I. 1961. Operational sequence diagrams in system design. *Human Factors* 3: 66–73.

Kyriakidis, M R, R Happee, and J C F de Winter. 2015. Public opinion on automated driving: Results of an international questionnaire among 5000 respondents. *Transportation Research Part F* 32: 127–140.

Langan-Fox, J, J M Canty, and M J Sankey. 2009. Human-automation teams and adaptable control for future air traffic management. *International Journal of Industrial Ergonomics* 39 (5): 894–903.

Lansdown, T C. 2002. Individual differences during driver secondary task performance: Verbal protocol and visual allocation findings. *Accident Analysis and Prevention* 34 (5): 655–662.

Larsson, A F L. 2012. Driver usage and understanding of adaptive cruise control. *Applied Ergonomics* 43(3): 501–506.

Larsson, A F L, K Kircher, and J A Hultgren. 2014. Learning from experience: Familiarity with ACC and responding to a cut-in situation in automation driving. *Transportation Research Part F* 27 (B): 229–237.

Lavie, T and J Meyer. 2010. Benefits and costs of adaptive user interfaces. *International Journal of Human-Computer Studies* 68: 508–524.

Lavrinc, D. 2014. This is how bad self-driving cars suck in the rain. http://jalopnik.com/this-is-how-bad-self-driving-cars-suck-in-the-rain-.

Lee, S W, J Park, A R Kim, and P H Seong. 2012. Measuring situation awareness of operation teams in NPPs using a verbal protocol analysis. *Analysis of Nuclear Energy* 43: 167–175.

Lee, J D and K A See. 2004. Trust in automation: Designing for appropriate reliance. *Human Factors* 46: 50–80.

Lees, M N and J D Lee. 2007. The influence of distraction and driving context on driver response to imperfect collision warning systems. *Ergonomics* 50 (8): 1264–1286.

Lenard, J, A Badea-Romero, and R Danton. 2014. Typical pedestrian accident scenarios for the development of autonomous emergency braking test protocols. *Accident Analysis and Prevention* 73: 73–80.

Lenard, J, D Russell, M Avery, A Weekes, D Zuby, and M Kuehn. 2011. Typical pedestrian accident scenarios for the testing of autonomous emergency braking systems. *Proceedings of the 22nd International Technical Conference on the Enhanced Safety of Vehicles*. Washington, DC, 13–16 June 2011.

Liu, Y and C D Wickens. 1994. Mental workload and cognitive task automaticity: An evaluation of subjective and time estimation metrics. *Ergonomics* 37: 1843–1854.

Ma, R and D B Kaber. 2005. Situation awareness and workload in driving while using adaptive cruise control and a cell phone. *International Journal of Industrial Ergonomics* 35: 939–953.

Madhaven, P and D A Wiegmann. 2007. Similarities and differences between human-human and human-automation trust: An integrative review. *Theoretical Issues in Ergonomics Science* 8 (4): 277–301.

Marques, J and C McCall. 2005. The application of inter-rater reliability as a solidification instrument in a phenomenological study. *The Qualitative Report* 10 (3): 439–462.

Matthews, G, S E Campbell, S Falconer, L A Joyner, J Huggins, K Gilliand, R Grier, and J S Warm. 2002. Fundamental dimensions of subjective state in performance settings: Task engagement, distress, and worry. *Emotion* 2: 315–340.

Matthews, G, J Szalma, A R Panganiban, C Neubauer, and J S Warm. 2013. Profiling task stress with the Dundee Stress State Questionnaire. In *Psychology of Stress*, by L Cavalcanti and S Azevedo (eds.), pp. 49–91. Hauppauge, NY: Nova Science Publishers.

McIlroy, R C, N A Stanton, and B Remington. 2012. Developing expertise in military communications planning: Do verbal reports change with experience? *Behaviour & Information Technology* 31 (6): 617–629.

Merat, N and A H Jamson. 2009. How do drivers behave in a highly automated car? *Proceedings of the 5th International Driving Symposium on Human Factors in Driver Assessment, Training and Vehicle Design*, Big Sky, MT, 22–25 June 2009. pp. 514–521.

Merat, N, A H Jamson, F C Lai, and O Carsten. 2012. Highly automated driving, secondary task performance, and driver state. *Human Factors* 54 (5): 762–771.

Merat, N, A H Jamson, F C Lai, M Daly, and O M Carsten. 2014. Transition to manual: Driver behaviour when resuming control from a highly automated vehicle. *Transportation Research Part F: Traffic Psychology and Behaviour* 27 (B): 274–282.

Mercedes. 2015. The long haul truck of the future. Accessed 10 May 2016. https://www.
    mercedes-benz.com/en/mercedes-benz/innovation/the-long-haul-truck-of-the-future/.

Merritt, S M, H Heimbaugh, J LaChapell, and D Lee. 2013. I trust it, but I don't know why:
    Effects of implicit attitudes toward automation on trust in an automated system. *Human
    Factors* 55: 520–534.

Michon, J A. 1985. A critical view of driver behaviour models: What do we know, what
    should we do? In *Human Behaviour and Traffic Safety*, by L Evans and R C Schwing
    (eds.), pp. 485–520. New York, NY: Plenum Press.

Miller, C A and R Parasuraman. 2007. Designing for flexible interaction between humans
    and automation: Delegation interfaces for supervisory control. *Human Factors* 49 (1):
    57–75.

Miller, M B, J D van Horn, G Wolford, T C Handy, M Valsangkar-Smyth, S Inati, S Grafton,
    and M S Gazzaniga. 2002. Extensive individual differences in brain activations dur-
    ing episodic retrieval are reliable over time. *Journal of Cognitive Neuroscience* 14 (8):
    1200–1214.

Moiser, K L and L J Skitka. 1996. Human decision-makers and automated design aids: Made
    for each other? In *Automation and Human Performance*, by R Parasuraman and M
    Mouloua (eds.), pp. 201–220. Mahwah, NJ: Erlbaum.

Montag, I and A L Comrey. 1987. Internality and externality as correlates of involvement in
    fatal driving accidents. *Journal of Applied Psychology* 72: 339–343.

Moroney, W F and M G Lilienthal. 2009. Human factors in simulation and training: An over-
    view. In *Human Factors in Simulation and Training*, by P Hancock, D Vincenzi, J Wise
    and M Mouloua (eds.), pp. 3–38. Boca Raton, FL: Taylor & Francis.

Mumaw, R J, N B Sarter, and C D Wickens. 2001. Analysis of pilots monitoring and perfor-
    mance on an automated flight deck. *Proceedings of the 11th International Symposium
    on Aviation Psychology*. Columbus, OH, 5–8 March 2001.

National Highway Traffic Safety Administration. 1997. A review of human factors studies
    on cellular telephone use while driving. Accessed 20 October 2015. www.nhtsa.gov/
    people/injury/research/wireless/c5.htm.

National Highway Traffic Safety Administration. 2013. *Preliminary Statement of Policy
    Concerning Automated Vehicles System*. Washington, DC: National Highway Traffic
    Safety Administration.

National Transportation Safety Board. 1994. Aircraft accident report: Stall and loss of control
    on final approach. NTSB-AAR-94/07, Washington, DC.

National Transportation Safety Board. 2010. Loss of control on approach Colgan Air, Inc.
    operating as continental connection flight 3407. NTSB-AAR-10-01, Washington, DC.

Neisser, U. 1967. *Cognitive Psychology*. New York, NY: Appleton Century Crofts.

Nilsson, L. 1993. Behavioural research in an advanced driving simulator – Experiences of the
    VTI system. *Proceedings of the Human Factors and Ergonomics Society 37th Annual
    Meeting*. Seattle, WA, 11–15 October 1993, pp. 612–616.

Nisbett, R E and T D Wilson. 1977. Telling more than we can know: Verbal reports on mental
    processes. *Psychological Review* 84 (3): 231–259.

Norman, D A. 1990. The problem with automation: Inappropriate feedback and interaction,
    not over-automation. *Philosophical Transactions of the Royal Society of London –
    Series B: Biological Sciences* 327 (1241): 585–593.

Noujoks, F, C Mai, and A Neukum. 2014. The effect of urgency of take-over requests dur-
    ing highly automated driving under distraction conditions. *Proceedings of the 5th
    International Conference on Applied Human Factors and Ergonomics*. Krakow,
    Poland, 19–23 July 2014, pp. 2099–2106.

Nowakowski, C, S E Shladover, and H S Tan. 2015. Heavy vehicle automation: Human factors
    lessons learned. *Procedia Manufacturing* 3: 2945–2952.

O'Hanlon, J F and M E McCauley. 1974. Motion sickness incidence as a function of the frequency and acceleration of vertical sinusoidal motion. *Aerospace Medicine* 45: 366–369.

Oron-Gilad, T and A Ronen. 2007. Road characteristics and driver fatigue: A simulator study. *Traffic Injury Prevention* 8 (3): 281–289.

Parasuraman, R. 2000. Designing automation for human use: Empirical studies and quantitative models. *Ergonomics* 43 (7): 931–951.

Parasuraman, R, P A Hancock, and O Olofinboba. 1997. Alarm effectiveness in driver-centred collision warning systems. *Ergonomics* 40: 390–399.

Parasuraman, R, R Molloy, and I L Singh. 1993. Performance consequences of automation-induced 'complacency'. *The International Journal of Aviation Psychology* 3: 1–23.

Parasuraman, R and V Riley. 1997. Humans and automation: Use, misuse, disuse, abuse. *Human Factors* 39 (2): 230–253.

Parasuraman, R, T B Sheridan, and C D Wickens. 2000. A model for types and levels of human interaction with automation. *IEEE Transactions on Systems, Man and Cybernetics Society* 30 (3): 286–297.

Parasuraman, R and C D Wickens. 2008. Humans: Still vital after all these years of automation. *Human Factors* 50: 511–520.

Patrick, J. 1992. *Training: Research and Practice*. London, UK: Academic Press.

Patten, C J D. 2013. Behavioural adaptation to in-vehicle intelligent transport systems. In *Behavioural Adaptation and Road Safety: Theory, Evidence and Action*, by C M Rudin-Brown and S L Jamson (eds.), pp. 161–176. Boca Raton, FL: CRC Press.

Patten, C J, A Kircher, J Östlund, L Nilsson, and O Svenson. 2006. Driver experience and cognitive workload in different traffic environments. *Accident Analysis and Prevention* 38 (5): 887–894.

Peng, Y, L Boyle, and S L Hallmark. 2013. Driver's lane keeping ability with eyes off road: Insights from a naturalistic study. *Accident Analysis and Prevention* 50: 628–634.

Pennington, N, R Nicolich, and J Rahm. 1995. Transfer of training between cognitive subskills: Is knowledge use specific? *Cognitive Psychology* 28: 175–224.

Perry, M. 2003. Distributed cognition. In *HCI Models, Theories and Frameworks*, by J M Carroll (ed.), pp. 93–224. San Francisco, CA: Morgan-Kaufmann.

Plant, K L and N A Stanton. 2016. Distributed cognition in search and rescue: Loosely coupled tasks and tightly coupled roles. *Ergonomics* 59 (10): 1353–1376.

Power, J D. 2012. US Automotive emerging technologies study results. http://autos.jdpower. com/content/press-release/gGOwCnW/2012-u-s-.

Price, J. 2016. Human factors in the design of traffic management systems. PhD thesis, Faculty of Engineering and the Environment, University of Southampton, Southampton, UK.

Putkonen, A and U Hyrkkänen. 2007. Ergonomists and usability engineers encounter test method dilemmas with virtual work environments. In *Engineering Psychology and Cognitive Ergonomics*, by D Harris (ed.), pp. 147–156. Berlin, Germany: Springer-Verlag.

Rasmussen, J, A M Pijtersen, and L P Goodstein. 1994. *Cognitive Systems Engineering*. New York, NY: Wiley.

Reinartz, S J and T R Gruppe. 1993. Information requirements to support operator-automatic cooperation. *Human Factors in Nuclear Safety Conference*. London, UK, 22–23 April 1993.

Retting, R A, S A Ferguson, and A T McCartt. 2003. A review of evidence-based traffic engineering measures designed to reduce pedestrian-motor vehicle crashes. *American Journal of Public Health* 93 (9): 1456–1463.

Revell, K M and N A Stanton. 2014. Case studies of mental models in home heat control: Searching for feedback, valve, timer and switch theories. *Applied Ergonomics* 45 (3): 363–378.

Richards, D and A Stedmon. 2016. To delegate or not to delegate: A review of control frameworks for autonomous cars. *Applied Ergonomics* 53 (B): 383–388.

Riley, V. 1994. A theory of operator reliance on automation. In *Human Performance in Automated Systems: Recent Research and Trends*, by M Mouloua and R Parasuraman (eds.), pp. 8–14. Hillsdale, NJ: Erlbaum.

Road Safety Analysis. 2013. Stepping out, pedestrian casualties: An analysis of the people and circumstances. Commissioned by the Parliamentary Advisory Council for Transport Safety. Accessed 24 June 2013. http://www.roadsafetyanalysis.org.

Rogers, Y. 1993. Coordinating computer-mediated work. *Computer-Supported Cooperative Work* 1: 295–315.

Rogers, Y. 1997. *A Brief Introduction to Distributed Cognition*. Brighton, UK: Interact Lab, University of Sussex.

Rolnick, A and R E Lubow. 1991. Why is the driver rarely motion sick? The role of controllability in motion sickness. *Ergonomics* 34: 867–879.

Rosen, E, J E Källhammer, D Eriksson, M Nentwich, R Fredriksson, and K Smith. 2010. Pedestrian injury mitigation by autonomous braking. *Accident Analysis and Prevention* 42 (6): 1949–1957.

Rouff, C and M Hinchey. 2012. *Experience from the DARPA urban challenge*. London, UK: Springer-Verlag.

Rudin-Brown, C M. 2010. 'Intelligent' in-vehicle intelligent transport systems: Limiting behavioural adaptation through adaptive design. *IET Intelligent Transport Systems* 4 (4): 252–261.

Rudin-Brown, C M and H A Parker. 2004. Behavioural adaptation to adaptive cruise control (ACC): Implications for preventive strategies. *Transportation Research Part F: Traffic Psychology and Behaviour* 7 (2): 59–76.

Russo, J E, E J Johnson, and D L Stephens. 1989. The validity of verbal protocols. *Memory & Cognition* 17: 759–769.

Ryan, B and C M Haslegrave. 2007. Developing a verbal protocol method for collecting and analysing reports of workers' thoughts during manual handling tasks. *Applied Ergonomics* 38: 805–819.

Saad, F and T Villame. 1996. Assessing new driving system systems: Contributions of an analysis of drivers' activity in real situations. *Proceedings of the 3rd Annual World Congress on Intelligent Transport Systems*. Orlando, FL, 14–18 October 1996.

Saffarian, M, J C F de Winter, and R Happee. 2012. Automated driving: Human factors issues and design solutions. *Proceedings of the Human Factors and Ergonomics Society 56th Annual Meeting*. Boston, MA, 22–26 October 2012, pp. 2296–2300.

Salas, E, C Prince, D P Baker, and L Shrestha. 1995. Situation awareness in team performance. *Human Factors* 37: 123–136.

Salmon, P M, M G Lenne, G H Walker, N A Stanton, and A Filtness. 2014. Using the Event Analysis of Systemic Teamwork (EAST) to explore the conflicts between different road user groups when making right hand turns at urban intersections. *Ergonomics* 57 (11): 1628–1642.

Salmon, P M, N A Stanton, and G H Walker. 2004. *National Grid Transco: Switching operations report*. Defence Technology Centre for Human Factors Integration Report.

Salmon, P M, N A Stanton, and K L Young. 2012. Situation awareness on the road: Review, theoretical and methodological issues, and future directions. *Theoretical Issues in Ergonomics Science* 13 (4): 472–492.

Salmon, P M, N A Stanton, G H Walker, and D P Jenkins. 2009. *Distributed Situation Awareness: Advances in Theory, Measurement and Application to Teamwork*. Aldershot, UK: Ashgate.

Salmon, P M, N A Stanton, G H Walker, C Baber, D P Jenkins, R McMaster, and M S Young. 2008. What really is going on? Review of situation awareness models for individuals and teams. *Theoretical Issues in Ergonomics Science* 9 (4): 297–323.

Salmon, P M, G H Walker, and N A Stanton. 2016. Pilot error versus sociotechnical systems failure: A distributed situation awareness analysis of Air France 447. *Theoretical Issues in Ergonomics Science* 17 (1): 64–79.

Sarter, N B. 2008. Investigating mode errors on automated flight decks: Illustrating the problem-driven, cumulative, and inter-disciplinary nature of human factors research. *Human Factors* 50 (3): 506–510.

Sarter, N B, R J Mummaw, and C D Wickens. 2007. Pilots' monitoring strategies and performance on automated flight decks: An empirical study combining behavioural and eye-tracking data. *Human Factors* 49 (3): 347–357.

Sarter, N B and D D Woods. 1995. How in the world did we ever get into that mode? Mode error and awareness in supervisory control. *Human Factors* 37: 5–19.

Sarter, N B, D D Woods, and C E Billings. 1997. Automation surprises. *Handbook of Human Factors and Ergonomics* 2: 1926–1943.

Searson, D J, R W G Anderson, and T P Hutchinson. 2014. Integrated assessment of pedestrian head impact protection in testing secondary safety and autonomous emergency braking. *Accident Analysis and Prevention* 63: 1–8.

Shelton, J and G Kumar. 2010. Comparison between auditory and visual simple reaction times. *Neuroscience & Medicine* 1 (1): 30–32.

Shepherd, A. 2000. *Hierarchical Task Analysis*. London, UK: Taylor & Francis.

Sheridan, T B. 1970. Big brother as driver: New demands and problems for the man at the wheel. *Human Factors* 12: 95–101.

Sheridan, T B. 1988. Trustworthiness of command and control systems. *Third IFAC /IFIP/ IEA/IFORS Conference on Analysis, Design and Evaluation of Man-Machine Systems*. Oulu, Finland, 14–16 June 1988, pp. 427–431.

Sheridan, T B and W Ferrell. 1974. *Man-Machine Systems: Information, Control, and Decision Models of Human Performance*. Cambridge, MA: MIT Press.

Sheridan, T B and W L Verplanck. 1978. *Human and Computer Control of Undersea Teleoperators*. Cambridge, MA: MIT Man-Machine Laboratory.

Sheridan, T B, C D Wickens, and R Parasuraman. 2008. Situation awareness, mental workload, and trust in automation: Viable, empirically supported cognitive engineering constructs. *Journal of Cognitive Engineering and Decision Making* 2 (2): 140–160.

Shorrock, S T and O Straeter. 2006. A framework for managing system disturbances and insights from air traffic management. *Ergonomics* 49 (12–13): 1326–1344.

Singleton, W T. 1989. *The Mind at Work: Psychological Ergonomics*. Cambridge, UK: Cambridge University Press.

Sivak, M and B Schoette. 2015. Motion sickness in self-driving vehicles. Report No. UMTRI-2015-12, University of Michigan Transport Research Institute.

SMART European Commission Study Report. 2010. SMART 2010/0064: Definition of necessary vehicle and infrastructure systems for automated driving, 1-111, version 1.2. Accessed 22 January 2015. http://vra-net.eu/news/smart-20100064-definition-of-necessary-vehicle-and-infrastructure-systems-for-automated-driving/.

Smiley, A and K A Brookhuis. 1987. Alcohol, drugs and traffic safety. In *Road Users and Traffic Safety*, by J A Rothengatter and R A de Bruin (eds.), pp. 83–105. Assen, The Netherlands: Van Gocum.

Smith, B W. 2013. SAE levels of driving automation. Accessed 3 April 2017. http://cyberlaw. stanford.edu/blog/2013/12/sae-levels-driving-automation.

Society for Automotive Engineers. 2016. Taxonomy and definitions for terms related to on-road motor vehicle automated driving systems. Accessed 12 December 2016. http:// standards.sae.org/j3016_201401/.

Sorensen, L J, N A Stanton, and A P Banks. 2011. Back to SA school: Contrasting three approaches to situation awareness in the cockpit. *Theoretical Issues in Ergonomics Science* 12 (6): 451–471.

Soualmi, B, C Sentouh, J C Popieul, and S Debernard. 2014. Automation-driver cooperative driving in presence of undetected obstacles. *Control Engineering Practice* 24 (1): 106–119.

Stanton, N A. 2014a. Representing distributed cognition in complex systems: How a submarine returns to periscope depth. *Ergonomics* 57 (3): 403–418.

Stanton, N A. 2014b. EAST: A method for investigating social, information and task networks. In *Contemporary Ergonomics and Human Factors 2014: Proceedings of the International Conference on Ergonomics & Human Factors*. Southampton, UK, 7–10 April 2014, pp. 395–402. Southampton, UK: CRC Press.

Stanton, N A. 2015. Responses to autonomous vehicles. *Ingenia* 62: 9.

Stanton, N A and C Baber. 2006. The ergonomics of command and control. *Ergonomics* 49 (12–13): 1131–1138.

Stanton, N A, P R G Chambers, and J Piggott. 2001. Situational awareness and safety. *Safety Science* 39: 189–204.

Stanton, N A, A Dunoyer, and A Leatherland. 2011. Detection of new in-path targets by drivers using stop and go adaptive cruise control. *Applied Ergonomics* 42 (4): 592–601.

Stanton, N A and P P Marsden. 1996. From fly-by-wire to drive-by-wire: Safety implications of automation in vehicles. *Safety Science* 24 (1): 35–49.

Stanton, N A and M Pinto. 2000. Behavioural compensation by drivers of a simulator when using a vision enhancement system. *Ergonomics* 43: 1359–1370.

Stanton, N A and P M Salmon. 2009. Human error taxonomies applied to driving: Generic driver error taxonomy and its implications for intelligent transport systems. *Safety Science* 47: 227–237.

Stanton, N A, P M Salmon, L Rafferty, G H Walker, C Baber, and D P Jenkins. 2013. *Human Factors Methods: A Practical Guide for Engineering and Design*. Aldershot, UK: Ashgate.

Stanton, N A, P M Salmon, and G H Walker. 2015. Let the reader decide: A paradigm shift for situation awareness in sociotechnical systems. *Journal of Cognitive Engineering and Decision Making* 9 (1): 44–50.

Stanton, N A, R Stewart, D Harris, R J Houghton, C Baber, R McMaster, P M Salmon et al. 2006. Distributed situation awareness in dynamic systems: Theoretical development and application of an ergonomics methodology. *Ergonomics* 49 (12–13): 1288–1311.

Stanton, N A, G H Walker, M S Young, T A Kazi, and P M Salmon. 2007b. Changing drivers' minds: The evaluation of an advanced driver coaching system. *Ergonomics* 50 (8): 1209–1234.

Stanton, N A and M S Young. 2005. Driver behaviour with adaptive cruise control. *Ergonomics* 48: 1294–1313.

Stanton, N A, M S Young, and B McCaulder. 1997. Drive-by-wire: The case of mental workload and the ability of the driver to reclaim control. *Safety Science* 27 (2–3): 149–159.

Stanton, N A, M S Young, and G H Walker. 2007a. The psychology of driving automation: A discussion with Professor Don Norman. *International Journal of Vehicle Design* 45 (3): 289.

Stewart, R, N A Stanton, D Harris, C Baber, P Salmon, M Mock, K Tatlock, L Wells, and A Kay. 2008. Distributed situation awareness in an Airborne Warning and Control System: Application of novel ergonomics methodology. *Cognition, Technology & Work* 10 (3): 221–229.

Strand, N, J Nilsson, M I Karlsson, and L Nilsson. 2011. Exploring end-user experiences: Self-perceived notions on use of adaptive cruise control systems. *Transport Systems IET* 5 (2): 134–140.

Strand, N, J Nilsson, I C Karlsson, and L Nilsson. 2014. Semi-automated highly automated driving in critical situations caused by automation failures. *Transportation Research Part F* 27 (B): 218–228.

Taylor, K and J P Dionne. 2000. Accessing problem-solving strategy knowledge: The complimentary use of concurrent verbal protocols and retrospective briefings. *Journal of Educational Psychology* 92 (3): 413–425.

Underwood, S E. 1992. *Delphi Forecast and Analysis of Intelligent Vehicle-Highway Systems through 1991: Delphi II.* Ann Arbor, MI: University of Michigan.

Underwood, G, D Crundall, and P Chapman. 2011. Driving simulator validation with hazard perception. *Transportation Research Part F* 14: 435–446.

Vagia, M, A A Transeth, and S A Fjerdingen. 2016. A literature review on the levels of automation during the years. What are the different taxonomies that have been proposed? *Applied Ergonomics* 53: 190–202.

Valero-Mora, P M, A Tontsch, R Welsh, A Morris, S Reed, K Touliou, and D Margaritis. 2013. Is naturalistic driving research possible with highly instrumented cars? Lessons learnt in three research centres. *Accident Analysis and Prevention* 58: 187–194.

van den Haak, M, M de Jong, and P Jan Schellens. 2003. Retrospective vs. concurrent think-aloud protocols: Testing the usability of an online library catalogue. *Behaviour & Information Technology* 22 (5): 339–351.

van der Laan, J D, A Heino, and D de Waard. 1997. A simple procedure for the assessment of acceptance of advanced transport telematics. *Transportation Research Part C: Emerging Technologies* 5: 1–10.

van der Veer, G C. 1993. Mental representations of computer languages – A lesson from practice. In *Cognitive Models and Intelligent Environments for Learning Programming*, by E Lemut, D Boulay and G Dettori (eds.), pp. 21–33. Berlin, Germany: Springer-Verlag.

van Gog, T, K A Ericsson, R M Rikers, and F Paas. 2005. Instructional design for advanced learners: Establishing connections between the theoretical frameworks of cognitive load and deliberate practice. *Education Technology Research and Development* 53 (3): 78–81.

van Gog, T, L Kester, F Nievelstein, B Giesbers, and F Paas. 2009. Uncovering cognitive processes: Different techniques that can contribute to cognitive load research and instruction. *Computers in Human Behaviour* 25: 325–331.

van Ratingen, M. 2012. Fifteen years for safer cars/AEB fitment survey. *Proceedings of the 15th Anniversary Event on Avoid the Crash with Autonomous Emergency Braking Systems*. Brussels, Belgium, 13 June 2012. Accessed 20 January 2015. http://www.ami-ando.com/FNPIBLY.html?page=760000.

Venkatesh, V, M Morris, G B Davis, and F D Davis. 2003. User acceptance of information technology: Toward a unified view. *MIS Quarterly* 27 (3): 425–478.

Vienna Convention. 1968. Legal instruments in the field of transport. Accessed 23 November 2015. https://www.unece.org/fileadmin/DAM/trans/conventn/crt1968e.pdf.

Vitalari, N P. 1985. Knowledge as a basis for expertise in systems analysis: An empirical study. *MIS Quarterly* 9 (3): 221–241.

Vollrath, M, S Schleicher, and C Gelau. 2011. The influence of cruise control and adaptive cruise control on driving behaviour – A driving simulator study. *Accident Analysis and Prevention* 43 (3): 1134–1139.

Walker, G H, H Gibson, N A Stanton, C Baber, P Salmon, and D Green. 2006. Event analysis of systematic teamwork (EAST): A novel integration of ergonomic methods to analyse C4i activity. *Ergonomics* 49 (12–13): 1345–1369.

Walker, G H, N A Stanton, C Baber, L Wells, H Gibson, P Salmon, and D Jenkins. 2010. From ethnography to the EAST method: A tractable approach for representing distributed cognition in Air Traffic Control. *Ergonomics* 53 (2): 184–197.

Walker, G H, N A Stanton, T A Kazi, P M Salmon, and D P Jenkins. 2009. Does advanced driver training improve situational awareness? *Applied Ergonomics* 40 (4): 678–687.

Walker, G H, N A Stanton, and P M Salmon. 2011. Cognitive compatibility of motorcyclists and car drivers. *Accident Analysis and Prevention* 43: 878–888.

Walker, G H, N A Stanton, and P M Salmon. 2015. *Human Factors in Automotive Engineering and Technology.* Aldershot, UK: Ashgate.

Walker, G H, N A Stanton, and P M Salmon. 2016. Trust in vehicle technology. *International Journal of Vehicle Design* 70 (2): 157–182.

Walker, G H, N A Stanton, L Wells, and H Gibson. 2005. *EAST methodology for air traffic control.* Defence Technology Centre for Human Factors Integration Report. Accessed 3 April 2017. https://www.defencehumancapability.com/.

Walker, G H, N A Stanton, and M S Young. 2001. Hierarchical task analysis of driving: A new research tool. In *Contemporary Ergonomics*, by M A Hanson (ed.), London: Taylor & Francis, pp. 435–440.

Walker, G H, N A Stanton, and M S Young. 2008. Feedback and driver situation awareness (SA): A comparison of SA measures and contexts. *Transportation Research Part F* 11: 282–299.

Wallace, D F, J J Winters, and J H Lackie. 2000. An improved operational sequence diagram methodology for use in system development. *Proceedings of the Human Factors and Ergonomics Society Annual Meeting* 44: 505.

Ward, N J. 2000. Automation of task processes: An example of intelligent transportation systems. *Human Factors and Ergonomics in Manufacturing* 10 (4): 395–408.

Weißgerber, T, D Damböck, M Kienle, and K Bengler. 2012. *Evaluation of a contact analogue head up display for driver assistance systems in highly automated driving.* Unpublished report, Technical University Munich, Munich, Germany.

Wertheim, A H. 1978. Explaining highway hypnosis: Experimental evidence for the role of eye movements. *Accident Analysis and Prevention* 10 (2): 111–129.

Weyer, J, D Fink, and F Adelt. 2015. Human-machine cooperation in smart cars: An empirical investigation of the loss-of-control thesis. *Safety Science* 72: 199–208.

Whyte, J IV, E Cormier, and R Pickett-Hauber. 2010. Cognitions associated with nurse performance: A comparison of concurrent and retrospective verbal reports of nurse performance in a simulated task environment. *International Journal of Nursing Studies* 47 (4): 446–451.

Wickens, C D and J G Hollands. 2000. *Engineering Psychology and Human Performance* (3rd ed.). Upper Saddle River, NJ: Prentice Hall.

Wiegmann, D A, A Rich, and H Zhang. 2001. Automated diagnostic aids: The effects of aid reliability on users' trust and reliance. *Theoretical Issues in Ergonomics Science* 2: 352–367.

Wilde, G J S. 1994. *Target Risk.* Toronto, Canada: PDE Publications.

Wilson, J R and J A Rajan. 1995. Human-machine interfaces for systems control. In *Evaluation of Human Work: A Practical Ergonomics Methodology*, by J R Wilson and E N Corlett (eds.), pp. 357–405. London, UK: Taylor & Francis.

Winner, H and S Hakuli. 2006. Conduct-by-wire following a new paradigm for driving into the future. *Proceedings of FISITA World Automotive Congress.* Yokohama, Japan, 22–27 October 2006.

World Health Organisation. 2004. World report on road traffic injury prevention. Accessed 27 June 2013. http://www.who.int/violence_injury_prevention/publications/road_traffic/world_report/en/index.html.

Wouters, P, F Paas, and J J G van Merrienboer. 2008. How to optimise learning from animated models? A review of guidelines based on cognitive load. *Review of Educational Research* 78: 645–675.

Young, M S, S A Birrell, and N A Stanton. 2011. Safe driving in a green world: A review of driver performance benchmarks and technologies to support 'smart' driving. *Applied Ergonomics* 42 (4): 529–532.

Young, M S and O Carsten. 2013. Designing for behavioural adaptation. In *Behavioural Adaptation and Road Safety: Theory, Evidence and Action*, by C M Rudin-Brown and S L Jamson (eds.), pp. 359–370. Boca Raton, FL: CRC Press.

Young, K L, P M Salmon, and M Cornelissen. 2012. Missing links? The effects of distraction on driver situation awareness. *Safety Science* 56: 36–43.

Young, M S and N A Stanton. 2002. Malleable attentional resources theory: A new explanation for the effects of mental underload on performance. *Human Factors* 44 (3): 365–375.

Young, M S and N A Stanton. 2004. Taking the load off: Investigations of how adaptive cruise control affects mental workload. *Ergonomics* 47 (8): 1014–1035.

Young, M S and N A Stanton. 2007a. Back to the future: Brake reaction times for manual and automated vehicles. *Ergonomics* 50 (1): 46–58.

Young, M S and N A Stanton. 2007b. What's skill got to do with it? Vehicle automation and driver mental workload. *Ergonomics* 50 (8): 1324–1339.

Young, M S, N A Stanton, and D Harris. 2007. Driving automation: Learning from aviation about design philosophies. *International Journal of Vehicle Design* 45 (3): 323–338.

Zand, D E. 1972. Trust and managerial problem solving. *Administrative Science Quarterly* 17 (2): 229–239.

Zheng, P and M McDonald. 2005. Manual vs. adaptive cruise control – Can driver's expectation be matched? *Transportation Research Part C* 13: 421–431.

Index